SCIENCE RESTORED

Jack Bosworth, PhD

Science Restored

ISBN 978-1-947514-03-4

Printed in the United States of America by

St. Clair Publications

PO Box 726

McMinnville, TN 37111-0726

http://stclairpublications.com/

Table of Contents

Dedication

I dedicate this book to Esther Bosworth. My mother was a loving and intelligent person who helped many people during her life. She tutored kids who couldn't read, she made clothes for people, she counseled those in trouble. At her memorial service it was said that she was love. She loved me always, she loved me unconditionally; when I was set upon by others, she would comfort me and make sure that I was OK. Her intelligence was sharp, but not caustic. I often sat in the kitchen with her and we talked at great length and in some depth about the nature of the world and about human psychology. Without her love and encouragement, I would not have accomplished as much, and this book would probably not exist.

Introduction

I was indoctrinated in the standard belief systems of mainstream science, just as most well-educated people of western civilization, but there's something in my makeup that doesn't allow beliefs to stick. Although I enjoy topics such as theoretical physics and abstract mathematics, I am fundamentally pragmatic in my approach to life. Thus, when I observe events that contradict part of my training, I do not disregard those experiences. I also do not discard my knowledge of science. In fact, I have found over and over that a scientific approach is the most effective way to learn about aspects of nature, even those not currently included in mainstream science.

For many years, I could not understand how so many people can ignore contradictory experience, especially scientists. The scientific creed is to investigate, understand and communicate. I am practical and scientific in my orientation, and history was not a field I had studied. When I finally decided to read about the origins of modern science, I was shocked to find that the answers to my questions were right there in plain sight. At the same time, I can understand how people can take the historical records at face value, and ignore the implications and contradictions.

The present work describes some aspects of my experience and outlines the history of the development of modern science, pointing out some of the inconsistencies in the historical records and showing what may be a more consistent interpretation of the events. The subsequent sections describe experiments that shed more light on the nature of the complex reality in which we live.

One more note before we begin—I see tremendous opportunities here, as many others have in the past. By publishing this set of methods, principles and experiments, I am opening those opportunities for you. I hope you find them interesting and useful.

Section I: The Problem

Chapter 1: A Contradiction

Have you ever wondered why police use psychics to solve difficult criminal cases? The very idea is fundamentally unscientific. And why is astrology so popular? The answer is the same—pragmatism—if it works, use it. Are psychic awareness and astrology real?

This book is about aspects of reality that were known to exist before modern science was born—and are ignored by mainstream science—for good reasons. Come with me on a journey through time and thought. We will discover these reasons, see why they no longer apply and begin a new chapter in science.

Our journey begins in the 1960's.

From age fourteen, I read and learned about science and the scientific method. Physics and mathematics were easy for me, and I expected to become a physicist, but I was interested in many areas of knowledge. My mother introduced me to Christianity, but science taught me that only the physical universe exists, so I dismissed religion as a means to avoid reality. I studied psychology, not in school, but as a lifelong endeavor to understand people—what they do and why they do it, their strengths and weaknesses, commonalities and differences, and ultimately, the internal structure of human consciousness. Extra-sensory perception, or ESP as it was called, seemed just as ridiculous as religion. I loved to debate and could argue decisively against religion or ESP with anyone—and often did.

My convictions were strong, but my only real belief system was the "scientific method." The scientific method is the foundation for the philosophy of science, the belief system and dogma that has brought us microwave ovens, penicillin, the atomic bomb, the Internet, and so much more.

In 1620 AD, Francis Bacon, the father of the scientific method, wrote, "There are and can exist but two ways of investigating and discovering truth. The one hurries on rapidly from the senses and particulars to the most general axioms, and from them, as principles and their supposed indisputable truth, derives and discovers the intermediate axioms. This is the way now in use. The other constructs its axioms from the senses and particulars, by ascending continually and gradually, till it finally arrives at the most general axioms, which is the true but unattempted way." [1]

Thus, according to Bacon, experience is the only basis for investigating and discovering truth. Apparently, most of the people of his day jumped to conclusions and then decided what axioms made their conclusions valid. I think his comment is ageless.

According to Bacon, the "true" way of using experience is to slowly develop an understanding from the detailed to the general. Consistent with Bacon's advice, the first step in the scientific method is "observation."

---

[1] Bacon, Francis. *Advancement of Learning and Novum Organum: With A Special Introduction by James Edward Creighton, PhD* (p. 316). Published in New York (1900) by the Colonial Press. Can be viewed online at http://www.biodiversitylibrary.org/item/59907#page/396/mode/1up

On that fateful day just before my 19th birthday, I had a summer job at Michigan State University. Currently a physics major at Lansing Community College, I was confident that I would transfer to the MSU physics department and complete a PhD in physics.

The day was warm and sunny as I approached the field. I looked around cautiously to be sure that they had moved all the bulls to another pasture. Those bulls were extremely expensive breeding stock, a ton a piece, and able to kill me with the flick of a head or the stomp of a foot.

The field was empty. I was safe. Carrying my pail of creosote and a six-inch paint brush, I walked across to the farthest bull shed, 500 yards from the barn. My primary task was to put a coat of creosote on the eight bull-sheds in the pasture, but farmers occasionally delivered loads of hay, and when a load arrived, I had to stop painting and help put the hay away in the barn.

Each bull shed was ten feet wide, twelve feet long. The roofs angled from seven feet down to six feet. The fronts being open, I would be covering the three outside walls with creosote.

By mid-morning, I had painted three of the bull sheds, and helped put away two loads of hay. I knew that several different farmers delivered the hay. I didn't know where they came from, and, if you know farmers, you know there was no regularity in the deliveries.

Two of the walls had a fresh coat of creosote, and the third was mostly painted, when I suddenly knew that if I didn't hurry, the next load of hay would arrive before I completed the wall. I *knew* it. This was something I had not experienced before. I was logical and always thinking. My thought processes were intimately familiar to me, and this "knowing" without physical evidence was new

I decided to act as if it were true. What difference would it make? If I completed the wall and no load of hay arrived, I would move on to the next bull shed. If the load of hay arrived as I "knew" it would, I would help put away the hay and *then* move to the next bull shed. I was mostly amused by this experience and wasn't taking it seriously. After all, it was just a hiccup in human psychology, and had no basis in fact.

Painting quickly, I was just putting on the last few brush strokes when I saw the wagon come over the hill heading toward the barn. My amusement turned serious. *I guess I was wrong. I'll have to look into it.*

I remember those thoughts clearly because it was the turning point of my life. Let's look at this experience, as I did, in terms of the scientific method. Remember that observation is the first step in the scientific method. I had made an observation—I had known something that I could not have known by physical means.

In Norman W. Edmund's version of the scientific method, stage two of the scientific method is, "Is there a problem?" [2]

Clearly, there was a problem. As Stephen Hawking put it, "Any physical theory is always provisional, in the sense that it is only a hypothesis: you can never prove it. No matter how many times the results of experiments agree

---

[2] *The Scientific Method* article located at Scientific Muse http://www.odycc.com/blog/the-scientific-method/

with some theory, you can never be sure that the next time the result will not contradict the theory. On the other hand, you can disprove a theory by finding even a single observation that disagrees with the predictions of the theory ... Each time new experiments are observed to agree with the predictions the theory survives, and our confidence in it is increased; but if ever a new observation is found to disagree, we have to abandon or modify the theory." [3]

I had made a new observation that disagreed with modern scientific theories. If the observation had been correct, I had to revise the theory. As a scientist, I was compelled to investigate. I had no idea what had happened or how to repeat the experience, but I knew that it was similar to the stories of some of the people with whom I had argued, and I hate misinformation, especially when I'm the one giving it. I also knew that further personal experience would be necessary in my investigations.

The scientific method is based on open-minded investigation. As Carl Sagan put it, "Scientists do not seek to impose their needs and wants on Nature, but instead humbly interrogate Nature and take seriously what they find." [4]

Then why is it that mainstream-scientists haven't noticed these phenomena? Was I the first scientist to have an experience that contradicts scientific theory as relates to physical causal factors? I'm not that egotistical. I'm confident that others have had similar experiences—and may have ignored them. And yet, according to Carl Sagan, "We are constantly prodding,

---

[3] Hawking, Stephen. (1998) *A Brief History of Time: From the Big Bang to Black Holes* (p. 10). New York, NY: Bantam Books.
[4] Sagan, Carl. (1996) *The Demon-Haunted World: Science as a Candle in the Dark* (p. 32). New York, NY: Random House.

challenging, seeking contradictions or small, persistent residual errors, proposing alternative explanations, encouraging heresy. We give the highest rewards to those who convincingly disprove beliefs." [5]

Really? I ask again, why is it that mainstream science has not investigated these phenomena? Do they see the investigation as being too difficult? Is that why they ignore it? Albert Einstein said, "I have little patience with scientists who take a board of wood, look for its thinnest part, and drill a great number of holes where drilling is easy." [6] Many scientists seek easy problems to investigate and solve, but I am also sure that other strong scientists have had similar experiences. So why is this not part of mainstream science?

Later on this journey, you will see evidence that approximately 500 years ago, the Christian Church and secular authorities forced scientists to ignore nonphysical reality, i.e. everything that isn't part of physical reality as we know it today. But, you argue, that was 500 years ago, and now it is acceptable for scientists to disagree with the church. Carl Sagan may have said it best, "One of the saddest lessons in history is this: If we've been bamboozled long enough, we tend to reject any evidence of the bamboozle. We're no longer interested in finding out the truth. The bamboozle has captured us. It is simply too painful to acknowledge—even to ourselves—that we've been so credulous." [7]

---

[5] Sagan, Carl. (1996) *The Demon-Haunted World: Science as a Candle in the Dark* (p. 32). New York, NY: Random House.
[6] Bynum, William F. and Porter, Roy (editors). (2017) *Oxford Dictionary of Scientific Quotations* (p. xx) Oxford, UK: Oxford University Press.
[7] Sagan, Carl. *The Fine Art of Baloney Detection* article. Parade Magazine (February 1, 1987). Can be viewed online at http://www.csicop.org/uploads/files/ParadeFeb11987.pdf

Although the church and governments still admonish and encourage scientists to ignore nonphysical reality, as Sagan suggested, today it appears that the scientists themselves choose *not* to see that they have been bamboozled. However, I think appearances may be deceiving, which I will explain, in more detail, in a later chapter. I know you are anxious to see how our science became limited, but first, I want to answer your question—"Is this guy crazy?"

Chapter 2: External Verification

The journey continues in 1971 ...

The notice proclaimed that classes would be held to teach extra-sensory perception. I seldom looked at bulletin boards, but this poster caught my eye. I remembered the commitment I had made a few years earlier to investigate whatever that experience had been in the field beside the bull shed. I was a graduate student at the University of Michigan, studying computer science. I had continued my studies in physics at Michigan State University until the end of my junior year, when I had switched to mathematics. At the time, it had seemed likely that more important advancements would be made in computer science than in physics, and, as a math honors student, I was allowed to take all the available computer science classes. Upon graduating from MSU, having specialized in abstract mathematics with lots of computer science, many graduate schools were open to me. I had chosen the U of M, and the PhD track in computer science.

The ESP classes were held at a local commune, and were not part of the university curriculum. Two men taught the classes, based on a course they had taken on "Silva Mind Control." [8] Interestingly enough, their intension was to develop talent among their students as a foundation for fundamental research in the field. To my knowledge, they were unable to get financing for their research, but they did teach us how to meditate and how to become

---

[8] Silva, José, and Miele, Philip. (1977) *The Silva Mind Control Method.* The complete text can be read online at https://archive.org/details/JoseSilvaTheSilvaMindControlMethod

aware, at least in limited ways, of nonphysical reality. Meditation and related techniques will be covered in later chapters.

During the ESP course, I found that I was not as talented as many of the other students—less able to perceive. Most of my perceptions were dark and vague. In fact, late in the course, that is why I chose to watch instead of participate in the most dramatic experiment we performed.

I was married to a woman who worked in the Intensive Care Unit (ICU) at the University Hospital. Normally, she gave me a ride home when she finished work. On this particular day, she arrived at the commune before we were ready to end the class. Another student stated the desire to "do one more case." A "case" was a special procedure wherein one person provided the name, age, sex and address of a person unknown to the person, or persons, who would "do the case." The participants would then "go down in levels," contact the named person, and begin to verbally share what they found. Our intention was to discover medical facts about the person; in previous cases, we had usually uncovered injuries or diseases that were verified by the person providing the "case" details.

I suggested that my wife provide the case. My motivation was two-fold. First, my wife had always been extremely skeptical about the ESP classes and thought that I was either crazy or stupid to attend, so I wanted her to view undeniable evidence of her own. Second, her case would be unknown to everyone in the class, and those doing the case, therefore, would have had no prior access to information about the target person.

She reluctantly provided the name, age, sex and address of a man, giving no other information, as per our instructions. Three of the more talented students in the ESP class lay down to do the case together. Each of them reported different things about the man in question. They reported tubes running into his body through his nose and mouth, broken bones, internal injuries to his organs, and more. When they were finished, my wife verified that everything they had said was accurate. The man was an accident victim who had been brought into the ICU earlier that day after extensive surgery. My wife never questioned my interest in ESP classes again.

One is forced to ask the same questions as in the previous chapter. Is it possible that no other scientist has ever witnessed such evidence as we saw in that ESP course?

Nah! If others have observed the same, have they chosen to ignore it because further investigation would be too difficult? I'm sure that would be true of some scientists, but here we had at least three effective research subjects. The research wouldn't be that difficult—or would it?

Peer-pressure is intense among scientists, as demonstrated in a side-trip we'll take a couple of years into the future. I presented a paper on function optimization at a conference at Princeton University. The popular approach to function optimization was called the "gradient method," which is based on calculus. As a research assistant, I had investigated the use of genetic algorithms to optimize functions in 40 dimensions. My results were radical, but based entirely on empirical evidence—results from a computer program that used a genetic algorithm to optimize actual 40-dimensional functions. The world leaders in function optimization research were in the audience when I presented my findings.

My sister attended my presentation, and she verifies that the other scientists were brutal and mean in their questioning. They even insinuated that I was lying. This critical attitude can be healthy in scientific research, but only when the new evidence is accepted after validation. I assume that those function optimization researchers did, indeed, check my data and then began to consider genetic algorithms.

However, scientists are human, and they exert great pressure on each other. Carl Sagan's words seem quite relevant. "Those who cannot bear the burden of science are free to ignore its precepts. But we cannot have science in bits and pieces, applying it where we feel safe and ignoring it where we feel threatened." [9]

The function optimization researchers in my audience were certainly threatened by my findings. If a new approach is that threatening in an established field of study, how threatening is a new field of study, especially when those scientists have been bamboozled into believing that the field is bogus?

Researchers need to question each other so they can separate the wheat from the chaff, the true from the misunderstood. But if the peer-pressure is too intense, it would certainly make truly new research difficult or impossible.

Let's continue the journey from where we left off. After completing the ESP course, I felt that I had a good basis for further investigation. I found a bookstore where a clerk recommended several books. As I read and

---

[9] Sagan, Carl. (1996) *The Demon-Haunted World: Science as a Candle in the Dark* (p. 297). New York, NY: Random House.

experimented with those books, I found a disturbing trend—all of them referenced astrology as if it were real and meaningful. Science told me that astrology was invalid, so I knew those authors had to be mistaken. After all, a test of astrology would be simple; surely other scientists had tested it thoroughly before deciding that it was meaningless.

Even so, I decided to run a simple experiment to prove that I was correct and that astrology was just a fun fantasy. My wife and I were taking a summer vacation in Montana and it would be a long ride in the car. Each of us thought of everyone whose astrological sun-sign we knew. There are twelve Zodiac signs—Aries, Taurus, Gemini, Cancer, Leo, Virgo, Libra, Scorpio, Sagittarius, Capricorn, Aquarius, and Pisces. The sun-sign is the Zodiac sign wherein the sun was located at the time of the person's birth. This can be calculated by exact astronomical means, and those astronomical calculations are the starting point of astrology.

Since the sun-sign is generally known by the month and day of a person's birth, we worked from each person's birthday. Between us, we listed 39 people and their birthdays. I had a simple sun-sign book, and for each person we read the descriptions of three sun-signs, only one of which was the sun-sign in which the person was born. My hypothesis that we would find the best fit was equally distributed among the three selected sun-signs. In terms of the scientific method, this is called the "null" hypothesis, i.e. that there would be no observable phenomenon—no correlation.

To my considerable surprise, I found that all but one of our subjects fit his or her sun-sign significantly better than either of the other two descriptions that we read. Admitting that I had been wrong, I was further committed to looking into the topic in more depth.

In scientific terms, what I had found was something called a direct correlation. I had hypothesized that there was no correlation between a person's sun-sign and that person's perceivable characteristics. Isn't that what mainstream scientists tell us? Instead of *no* correlation, I had found a *strong*, direct correlation.

Carl Sagan wrote, "No stuffy dismissal by a gaggle of scientists makes contact with the social needs that astrology—no matter how invalid it is—addresses, and science does not." [10] Apparently, Sagan believed astrology to be invalid. I don't think that Carl Sagan would have lied about it, so I assume that he never ran a test of astrology such as mine or those used by C.G. Jung. [11] It probably didn't occur to him. The bamboozle caught him, too.

Again, the same questions—have no other scientists looked into astrology? Of *course*, some have! Then why do so many scientists believe it is invalid? Why don't they know about the correlations that exist and investigate them? The answers will be developed later on this journey, but first, let's consider a few other relevant topics.

---

[10] Sagan, Carl. (1996) *The Demon-Haunted World: Science as a Candle in the Dark* (p. 304). New York, NY: Random House.
[11] Roderick, Main. (1998) *Jung on Synchronicity and the Paranormal*. Princeton, NJ: Princeton University Press.

Chapter 3: More Possibilities

Continuing the journey a few years after I finished graduate school, I was working in Columbus, Ohio. I had been reading about phenomena called "power spots" and wanted to experience one for myself. My wife and a friend wanted to participate. We had decided to go to Colorado for a "working" vacation. I was sure that Colorado would have power spots, whatever those were, and it would be a nice place to hike and climb. In preparation, I sat in my living room and astral traveled (a type of controlled out-of-body experience) to Colorado. Not sure what to look for, I simply watched for something interesting while traveling around the mountains on the astral level of reality (LOR). (The nonphysical LOR closest to the physical is called the "etheric" level. The "astral" LOR is immediately above the etheric level. By way of reference, the human aura resides on the etheric LOR. More about this later ...) I saw what appeared to be a beam of light shining from the side of a mountain. Thinking that it might be a power spot, I made note of landmarks and returned to my physical body in Columbus. We had a Colorado map spread out on the floor, and I marked the map at the point of origination of the beam. The map was not detailed, so the mark could have been off by as much as a couple of miles, but I described the physical terrain so that we would recognize it when we got there.

A week later, we drove to Colorado. When we arrived in the vicinity of the marked spot, we stopped on a nearby road and I again astral traveled to find the spot. I found it roughly five miles to the north and slightly west of our location. By then, we had a detailed topographic map of the immediate area, and I carefully marked the spot on the map. I wanted to triangulate the spot to be sure exactly where it was before we left the car, so we went through the same procedure on two other nearby roads. All three times, my estimation of the location of the spot was the same, so we found a convenient place to leave the car and began to hike across the low mountain. Using the topo map and a compass, we were moving among ten-foot pine trees when we came upon four forest rangers. They were cutting trees that had been infected by the Mountain Pine Beetle. I stopped to ask one of the rangers about the spot we were seeking. Needless to say, I did not explain what we expected to find. I only described the terrain, which was a small saddle-back formation backed by a rock wall. The ranger stated that no such physical formation existed on that side of the mountain, and returned to his work.

You may have noticed that I tend to be persistent in the face of discouraging words. I knew from our map that we were only a quarter of a mile from the spot, and we continued to it ... where we found the exact formation I had described to the ranger.

The point of the story is that I was able to perceive physical formations from over a thousand miles away.

Did we find a power spot? Yes, about 40 miles from that first location, we found a large and intense power spot. Etheric-level energy appears to emanate from the ground at the base of a power spot. That is an observable aspect of a power spot—for those who have etheric vision.

A few years after that Colorado trip, I was visiting a place called Old Man's Cave, a little south of Columbus, Ohio. I had discovered a power spot there and was making further observations. That power spot is interesting in that the energy comes out of the ground at a 45 degree angle and passes over a frequently used paved walking trail, some forty feet away. Taking a break from my observations of the power spot, I was sitting on a rock twenty feet to the east. I was surprised by the number of people who stopped directly under the energy beam. The spot was not marked physically and the view from there was not otherwise interesting. Each person or group of people who stopped there seemed to look around as if they wondered why they had stopped.

A family came along the trail and stopped under the energy stream. The father, mother and eight-year-old boy all looked around just like everyone else. But the four-year-old boy looked directly at the base of the energy beam, tilting his head to follow the beam up and over his head. Then he looked at his older brother and his parents, and began to look around like they were. Maybe we learn young how (not) to perceive the world.

Our journey continues eight years later. I was living in Colorado Springs. I enjoy scaling the 14-thousand-foot mountains in Colorado, and on that particular day, I had awakened long before dawn with the intention of climbing El Diente. Parking the car, I began to hike north in the predawn moonlight. The trail led through a forest, then turned right near the edge of the forest. Hours later, I was unable to find a route to the summit and gave up. By mid-afternoon, I was on my way back to the car. I decided to take a shortcut, and left the trail, heading through the forest. (I've found that shortcuts are usually a bad idea in such situations.) When I had hiked for a while through the woods, I realized that I was lost. I remembered that the trail was too faint to notice, even if I were stepping across it, so I had little hope of finding it again I was considering my options when I remembered that I could astral travel. I sat down, went into meditation, and left my body, moving a little above the forest. I could see my car in the distance. I marked the direction, and, after returning to my body, checked it on my compass. Following that compass heading, I walked to the edge of the forest. I could see my car a half mile away—exactly on that heading.

At this point, you're asking yourself, "Where are we going on this journey?" The point of these excursions into my experience is to be suggestive. What if these experiences were common-place? How would the world be different? What if forest rangers were trained in the technique, i.e. they could astral travel? Hikers would be more secure in the mountains because the rangers could check on them periodically.

Police already use psychics to support them in solving the most difficult crimes. What if a special unit of the police were trained to astral travel? Imagine how much easier it would be for them to shield us from bad people. We could walk the streets at all hours with increased confidence.

What if your doctor had been trained to do Silva "cases" like in that ESP course? Imagine how much easier it would be for us to stay healthy. He would, for example, be able to see the initial formation of a cancer before you knew anything was wrong. You would want to visit your doctor a few times each year to stay in tip-top shape.

These are significant applications of etheric vision and astral travel, but there are so many more. I think the quality of our lives could be greatly enhanced, we could live longer, and our technology could grow in many ways. Imagine, for example, having more understanding and empathy for, and from, those we love. With increased mutual understanding, peace could grow, at home and everywhere. Even a little more peace would be good for the world.

You may ask, "What about resulting technological changes?" While I think it is too soon to speculate, the possibilities are endless. Until we investigate the nature of other LORs and how they interact with the physical level and with each other, we have little basis for guesses.

However, we will see later in this journey that some physical mechanisms appear to influence the etheric level and some etheric events seem to influence the physical level. We will also see that the natural laws are probably not the same on other LORs. We need to investigate and develop an understanding of the natural laws on each level, as scientists have done on the physical level. I do know a few things that are suggestive. I've noticed that travel on the astral level is much faster than the speed of light. I've also seen that energy can be translated from one LOR to another. I've read that matter can also be translated from one level to another. Putting these together, if we could translate physical matter to the etheric (and maybe to the astral) LOR ... practical inter-stellar travel. In keeping with the Information Age, I wonder what would happen if computers and communication were not limited by the speed of light.

All right, onward, but keep in mind that no one taught me how to do the things I'm reporting to you in this book (except the Silva "cases"). I've learned techniques by reading, experimenting and practicing. The experiences I'm reporting are not unique, or even truly unusual, and I will provide enough references so that you can verify all of this for yourself. My point is that even though I am not as talented as many who can perceive as I do, my awareness is very useful. What if my level of awareness were common among scientists? Even more to the point, why isn't it? Let's find out ...

Section II: An Historical Perspective

Chapter 4: Pre-Renaissance

Let's go back to ancient times to take a look at science and the prevailing attitudes. My purpose is to paint a picture of society, focusing on those parts of the painting that affected the development of science.

What was known, say 30,000 years ago? The Sun and Moon rise and set. The stars come out at night. The seasons change. When hungry, one eats plants or finds an animal and kills it. When lustful, one finds one of the opposite sex and ... they probably didn't know why women had babies, but I'm sure they enjoyed the sex and they cared for the babies, none the less. People didn't live long. In short, the world was a dangerous and mysterious place. People wanted to understand and gain control over their existence—just as we do today.

The Encylopaedia Britannica states, "In general, a science involves a pursuit of knowledge covering general truths or the operations of fundamental laws." [12] It further describes the history of science as follows.

"On the simplest level, science is knowledge of the world of nature. There are many regularities in nature that mankind has had to recognize for survival since the emergence of *Homo sapiens* as a species. The Sun and the Moon periodically repeat their movements. Some motions, like the daily motion of the Sun, are simple to observe; others, like the annual motion of the Sun, are far more difficult. Both motions correlate with important terrestrial events.

---

[12] *Science* entry retrieved July 6, 2006, from Encyclopædia Britannica 2006 Ultimate Reference Suite DVD.

Day and night provide the basic rhythm of human existence; the seasons determine the migration of animals upon which humans depended for millennia for survival. With the invention of agriculture, the seasons became even more crucial, for failure to recognize the proper time for planting could lead to starvation. Science defined simply as knowledge of natural processes is universal among mankind, and it has existed since the dawn of human existence." [13]

Was man that logical and pragmatic? No, of course not, and people *needed* to understand why—why do the seasons change, why does a woman have a baby, why do plants grow, why does a stream flow—why? Gods were postulated to control everything. Even today, most humans believe in one or more gods. What role do gods really play in the world?

"... since the deity or deities were themselves rational, or bound by rational principles, it was possible for humans to uncover the rational order of the world. This combination of religion and astronomy was fundamental to the early history of science. It is found in Mesopotamia, Egypt, China (although to a much lesser extent than elsewhere), Central America and India. The spectacle of the heavens, with the clearly discernible order and regularity of most heavenly bodies highlighted by extraordinary events such as comets and novae and the peculiar motions of the planets, obviously was an irresistible intellectual puzzle to early mankind. In its search for order and regularity, the human mind could do no better than to seize upon the heavens

---

[13] *History of Science* entry retrieved July 6, 2006, from Encyclopædia Britannica 2006 Ultimate Reference Suite DVD.

as the paradigm of certain knowledge. Astronomy was to remain the queen of the sciences (welded solidly to theology) for the next 4,000 years." [14]

Not everyone has held the opinion that God or the gods are rational or bound by rational principles. Consider your fundamental biases today. Living in the 3rd millennium AD, we have grown up with a science that is based on "natural laws." We depend on computers wherein even the tiniest bits of stable matter are assumed to be completely predictable (with the exception of quantum theory, which is as yet ignored by most people). What if God occasionally chose to have the electrons run backward in the circuit? Or maybe He would sometimes choose to have 2+2 add up to 5. Our use of computers, cars, cell phones, etc. all require nature to be dependable. So what if gods could change nature on a whim? Or, more importantly, what if you *believed* that gods controlled everything and that those gods do change things in the world around you based on their moods and whims?

That was the case for many (most) people prior to the development of modern science—they believed that the perceived reality was only stable as long as the gods wished it to be. [15] [16] Some believed that today the gods placed Venus there in the sky, but tomorrow, the planet might be placed somewhere else—or removed completely. According to *A History of the Ancient World*, "The number of divine elements in man's life was

---

[14] *History of Science* entry retrieved July 6, 2006, from Encyclopædia Britannica 2006 Ultimate Reference Suite DVD.
[15] Steele, Philip. (2005) *Galileo: The Genius Who Faced The Inquisition*. Washington, D.C.: National Geographic.
[16] Frazer, Sir James George *The Golden Bough: A Study of Magic and Religion*. Can be read online at http://www.templeofearth.com/books/goldenbough.pdf

unlimited."[17] Thus, it was common practice to perform rituals and make sacrifices to the gods to gain their favors—to have more male children, to be healthy, to win in battle, to have a good harvest, and anything else a man or woman might wish for. Even today, people pray to God for these very same things, hoping that He will grant them miracles.

Now consider the effect of these beliefs on the development of science. Imagine, if you can, that you are living in 1000 BC. You believe in gods who bring the seasons, grow trees, bring the night and day, move the planets and stars in the sky, bring life and death, etc. Suppose you observed the regularity of the motions of the planets and knew some mathematics. Could you conclude that the planets always move in roughly circular orbits around the sun? Probably not. Your most likely conclusion would be that the gods were happy. If they weren't happy, the gods would probably remove the planets from the skies, they would bring on a permanent winter, and the sun would not rise.

With such a belief system, science could not develop. History tells us of the occasional exceptional men who applied logic in spite of the beliefs of the time, such as Hippocrates, Plato, Aristotle and Ptolemy. Although these men were influential in their respective times, their use of observation and logic could not overcome the preponderance of beliefs in fickle gods. Remember that today, some people still believe that the Earth was created about 6000 years ago, even in the face of evidence of manmade tools in India—

---

[17] Rostovtzeff, M. (1927) *A History of the Ancient World: Volume I: The Orient and Greece*. Oxford, UK: The Clarendon Press. Translated from the Russian by J. D. Duff (1971); Greenwood Press.

"...quartzite pebble tools and flakes date to about two million years ago..." [18] Consider how much of known science these people have had to ignore to maintain their belief in the short existence of the Earth. I do not think they would be able to investigate and *develop* the portions of science their beliefs forced them to ignore.

Let's go back to the beliefs prevalent in ancient times. History shows us that people turned to wizards and priests for help in all sorts of matters. Some of the Magi claimed to understand the gods and even to be able to control the gods to some extent. This is important to keep in mind. Humans have a need to understand *and control* their lives. Astrology and belief in the gods provided some degree of understanding of events and circumstances. Rituals and sacrifices to the gods provided only a minimal sense of control. Wizards and priests gave a greater sense of control. Today, if your computer fails to function, you call a computer technician ... and you say that what he or she does is magic, because you don't understand. 2,000 years ago, you were simply called a different type of wizard.

Astrology was applied to understand and/or predict all aspects of life. For example, "In later centuries of Imperial China, it was universal practice to have a horoscope cast for each newborn child and at all decisive junctures in life." [19]

Medicine is another example. Disease was understood primarily in terms of superstition until Hippocrates (circa 460—circa 370 BC), around 400 BC,

---

[18] *India* entry retrieved July 7, 2006, from Encyclopædia Britannica 2006 Ultimate Reference Suite DVD.
[19] *Astrology* entry from Volume 1 (2005) Micropaedia, p.654.

developed an empirical approach to medicine. Attributed as having voiced the words, "Science is the father of knowledge, but opinion breeds ignorance," [20] Hippocrates is regarded as the father of modern medicine. He may also have been one of the earliest people to propose the basic principles of the current scientific method, stating, "... I approve of theorizing also if it lays its foundation in incident, and deduces its conclusions in accordance with phenomena." [21] Hippocrates also said, "He who does not understand astrology is not a doctor, but a fool." [22]

Was he just ignorant, or did Hippocrates see some useful correlations between human health and the movements of the heavenly bodies? While we will revisit the validity of astrology many times in this book, I must point out a common misconception. The Encylopaedia defines astrology as a "type of divination that involves the forecasting of earthly and human events through the observation and interpretation of the fixed stars, the Sun, the Moon, and the planets." [23] Although astrology is often used to predict the future, Isabel Hickey, a recognized authority, on astrology, wrote, "The stars impel but do not compel. An understanding of planetary influences allows you to take your life into your own hands and intelligently utilize the planetary influences ... " [24] In other words, the astrological influences no more

---

[20] Retrieved July 4, 2006 from https://en.wikiquote.org/wiki/Talk:Hippocrates
[21] Jones, W. H. S. (editor). (1968) *Hippocrates Collected Works*. Cambridge, MA: Cambridge Harvard University Press. Retrieved July 5, 2006 from the Digital Hippocrates Collection located at http://www.chlt.org/sandbox/dh/index.html
[22] Retrieved July 4, 2006 from https://en.wikiquote.org/wiki/Talk:Hippocrates
[23] *Astrology* entry retrieved July 4, 2006, from Encyclopædia Britannica 2006 Ultimate Reference Suite DVD.
[24] Hickey, Isabel M. (1992) *Astrology: A Cosmic Science: The Classic Work on Spiritual Astrology* (located in the preface). CRCS Publications.

determine a person's future than do the ocean currents determine the course of a ship. But by the same token, no sea captain ignores the currents.

As pragmatic as Hippocrates was in his teachings, it seems likely that he found a practical application for astrology in medicine, even in the absence of accurate astronomical data. Astrology was also applied in many other areas of human life, such as military operations, individual decisions/issues, when to plant or harvest crops, etc.

So where are we, as we approach the Christian era? Various forms of astrology and magic have developed in many parts of the world in attempts to understand and control the influences on human existence. One or more gods are believed to control the Earth and the heavens—everything.

Christianity began to spread in the latter half of the first century AD, and the Christian Bibles began to take their current forms near the end of the 4th century AD. [25] Many versions of the Bible have been written, but they are relatively consistent as they relate to the teachings of Jesus Christ. According to the Encylopaedia, "All the acts of a disciple <of Christ> must express love and forgiveness, even to enemies..." [26] I. e. the religion is based on love, rather than violence.

Thus, it may seem surprising that the Christian Bible includes Exodus 22:18: "Thou shalt not suffer a witch to live." [27] This is a direct command to all

---

[25] *Council of Nicea* entry retrieved July 8, 2006, from Encyclopædia Britannica 2006 Ultimate Reference Suite DVD.
[26] *Christianity* entry retrieved July 8, 2006, from Encyclopædia Britannica 2006 Ultimate Reference Suite DVD.
[27] The Bible (King James Version), Exodus 22:18.

believers to kill any witch they might run across, and to think that the passage is still there today. Wow! This command has also been translated to mean that you should kill wizards, alchemists, magicians (generally not including slight-of-hand) and necromancers. In fact, in 385 AD, "One heretic, Priscillian, was even executed for witchcraft (385), but in the face of vehement church protests." [28]

The church was attempting to consolidate and strengthen its position of power in the world, but at least at that time, it did not condone the killing of witches. So why was Priscillian killed for practicing witchcraft? We can only surmise. Let's consider the beliefs of the time. A witch was believed to be able to "bewitch" another person, causing unpleasant events to occur in the victim's life. Only a wizard or witch (or someone else aware in the right ways) could determine if an unfortunate event was natural or had been caused by witchcraft. As the Britannica puts it, "Those who interpret their misfortunes in terms of witchcraft will often use similar means to discover the source of their woes..." [29]

This suggests that most people were no more aware of nonphysical reality than they are today. Imagine that you are living a few hundred years AD. You know nothing of science—probably not even astrology. You believe in a God who can change anything, at any time, and you know much about the old gods and devils of the woods, the water, etc. Your life is already tenuous,

---

[28] *Christianity* entry retrieved July 9, 2006, from Encyclopædia Britannica 2006 Ultimate Reference Suite DVD.
[29] *Witchcraft* entry retrieved July 9, 2006, from Encyclopædia Britannica 2006 Ultimate Reference Suite DVD.

and you believe that there are people who can influence your life for good or ill, whenever they so choose. There can be little doubt that you fear witches.

People were afraid of witches, whether it was justified or not. This fear continued through to the Middle Ages. Were these beliefs wrong or were they realistic? The church supports these beliefs while modern science and most modern governments deny them. I suggest you keep an open mind and apply the scientific method. Many of the beliefs, and much of the fear, were probably unwarranted, but only through objective investigation can we determine the truth.

At this point in our journey, at the end of the Middle Ages, logic is not as important as beliefs. Our next stop is the Renaissance—beautiful art, the birth of modern science, and—be careful to keep your hands safely inside the vehicles, as torture and death are common in these times for those who believe as we do.

Chapter 5: The Renaissance and the Inquisition

One of the factors that brought an end to the Middle Ages was, ". . . during the 12th century the cultivation of beans made a balanced diet available to all social classes for the first time in history. The population therefore rapidly expanded, a factor that eventually led to the breakup of the old feudal structures." [30] Beans! I can just imagine the atmosphere of the times, but seriously, as simple as it sounds, an inexpensive source of good nutrition—or lack thereof—has brought about many changes throughout history.

Our journey continues at the beginning of the Renaissance, but don't forget your belief system from a thousand years earlier. You will see that common beliefs during the Renaissance—when the foundation was laid for modern science—had not changed much.

The Encyclopaedia defines the Renaissance as "literally 'rebirth,' the period in European civilization immediately following the Middle Ages, conventionally held to have been characterized by a surge of interest in classical learning and values. The Renaissance also witnessed the discovery and exploration of new continents, the substitution of the Copernican for the Ptolemaic system of astronomy, the decline of the feudal system and the growth of commerce, and the invention or application of such potentially powerful innovations as paper, printing, the mariner's compass, and gunpowder. To the scholars and thinkers of the day, however, it was

---

[30] *Middle Ages* entry retrieved July 10, 2006, from Encyclopædia Britannica 2006 Ultimate Reference Suite DVD.

primarily a time of the revival of classical learning and wisdom after a long period of cultural decline and stagnation."[31]

Strangely enough, the beginning of the Renaissance coincided with the beginning of the Inquisition. "With the appearance of large-scale heresies in the 11th and 13th centuries—notably among the Cathari and Waldenses—Pope Gregory IX in 1231 instituted the papal Inquisition for the apprehension and trial of heretics. The use of torture to obtain confessions and the names of other heretics was at first rejected but was authorized in 1252 by Innocent IV."[32] Although this was the official beginning of the Inquisition, only a few "... people in the 13th and 14th centuries are burned, accused of necromancy or witchcraft or consorting with the Devil..."[33] I.e. not many practitioners of these systems were killed until later years.

Astrology, alchemy, necromancy, magic, and especially witchcraft, were considered to be heresy. However, the use of astrology and the various forms of magic were so common that the church did not initially attempt to control them except in extreme cases. In fact, "There was no European court or indeed no papal court in the 15th and 16th centuries that did not have an astrologer, that did not have a magus."[34] Thus, even though astrology and

---

[31] *Renaissance* entry retrieved July 10, 2006, from Encyclopædia Britannica 2006 Ultimate Reference Suite DVD.
[32] *Inquisition* entry retrieved July 9, 2006, from Encyclopædia Britannica 2006 Ultimate Reference Suite DVD.
[33] Professor Teofilo F. Ruiz. Professor at the University of California at Los Angeles. Terror of History: Mystics, Heretics, and Witches in the Western Tradition course http://www.thegreatcourses.com/courses/terror-of-history-mystics-heretics-and-witches-in-the-western-tradition.html
[34] Ibid.

magic were not officially condoned; the church and secular authorities commonly used their services—much as they use lawyers today.

As I mentioned above, few were killed for the crime of heresy before the 15th century, but the death-toll rose quickly. From a translation of The *Malleus Maleficarum*—"Estimates of the death toll during the Inquisition worldwide range from 600,000 to as high as 9,000,000 (over its 250 year long course) ..." [35] The *Malleus Maleficarum* was, of course, referring to the *Witch Craze*, wherein people were accused, convicted, and killed as witches. This accounted for most of the deaths of the Inquisition. According to Professor Ruiz, during the *Witch Craze* "between 80,000 and 100,000 people were executed in Western Europe, the majority of them older women, between the late 15th century and the 1660s... " [36]

Even if we believe the smaller numbers, at least 80,000 people were killed after being accused as witches. What changed? Why did the Inquisition become so extreme? The answers are complex and debatable.

First, and most important, the church was growing weaker. Dion Fortune, one of the foremost occultists of the 20th century, wrote, "It was not until the 15th century, when the power of the church was beginning to show signs of weakening, that men dared to commit to paper the traditional wisdom of

---

[35] *The Malleus Maleficarum* (1486) by Catholic clergyman Heinrich Kramer. Transcribed by Wicasta Lovelace and Christie Rice. Can be viewed online at http://www.malleusmaleficarum.org/
http://bibliotecapleyades.lege.net/cienciareal/cienciareal12.htm

[36] Professor Teofilo F. Ruiz. Professor at the University of California at Los Angeles. Terror of History: Mystics, Heretics, and Witches in the Western Tradition course http://www.thegreatcourses.com/courses/terror-of-history-mystics-heretics-and-witches-in-the-western-tradition.html

Israel."[37] A current web site discusses the decline of the Roman Catholic Church of the Renaissance in some detail, mentioning "humiliation of the papacy" by the French king in the early 14th century, the Great Schism, and saying, "Instead of providing spiritual direction in a rapidly changing world, the papal court was preoccupied with the development of its administrative machinery and with the collection of revenue."[38] Basically, the church was divided and in turmoil. Many church leaders were more concerned about politics and wealth than about maintaining the position of the church. Even the apostolic nature of its leadership was in question.

As mentioned earlier, the feudal system was also in trouble.[39] The population was increasing and the economic structure was changing.[40] The remaining rulers needed to establish and maintain control over their domains.[41]

Beggars became a threat to the land owners, especially elderly women who could not work for a living. To this economic threat was added fear. If you turned an old woman away from your door and she said, "Curse you!," was she a witch? Were you in danger of reprisal? Remember your belief that a witch could harm you without your knowledge. If that old woman was a

---

[37] Fortune, Dion. (2000) *The Mystical Qabalah* (p. 6). Newburyport, MA: Red Wheel Weiser Conari LLC. This is a revised edition.
[38] Renaissance.MSN Encarta. Retrieved July 23, 2006 from http://encarta.msn.com/encyclopedia_761554186_4/Renaissance.html
[39] *Middle Ages* entry retrieved July 10, 2006, from Encyclopædia Britannica 2006 Ultimate Reference Suite DVD.
[40] Professor Teofilo F. Ruiz. Professor at the University of California at Los Angeles. Terror of History: Mystics, Heretics, and Witches in the Western Tradition course http://www.thegreatcourses.com/courses/terror-of-history-mystics-heretics-and-witches-in-the-western-tradition.html
[41] Ibid.

witch, your next child might be still-born, or you might be injured by a horse and become an invalid.

The weakened church became further threatened by other events. In 1453, Constantinople fell to the Turks [42] and the displaced Greeks took their literature to Europe, mostly to Italy. [43] The Greek literature included works on astrology, alchemy, magic, and Hermeticism. So why was this a threat to the church? Remember your belief system in those times. You believed that one or more gods caused everything to be as it was, and that the gods could change anything at any time. And according to Professor Ruiz, the scholars believed that true knowledge lay in the distant past, in "deep time," and not in the present or future. Thus, if you could access knowledge from the beginning of time, you would know the truth. [44] And the kicker? They believed that Hermes Trismegistus, the author of the Hermetic documents, had been a contemporary of Moses. [45] (Today, historians believe that these writings actually originated early in the Christian era. [46])

Do you see the problem for the church? Hermeticism was considered by the scholars to be in direct competition with the Bible! The foundation of the church was shaking.

---

[42] *Greek Literature* entry retrieved July 25, 2006, from Encyclopædia Britannica 2006 Ultimate Reference Suite DVD.
[43] Professor Teofilo F. Ruiz. Professor at the University of California at Los Angeles. Terror of History: Mystics, Heretics, and Witches in the Western Tradition course http://www.thegreatcourses.com/courses/terror-of-history-mystics-heretics-and-witches-in-the-western-tradition.html
[44] Ibid.
[45] Ibid.
[46] *Hermetic Writings* entry retrieved July 25, 2006, from Encyclopædia Britannica 2006 Ultimate Reference Suite DVD.

All right, so the church was threatened by this newly introduced Greek literature, but how bad was it, really? Books had to be copied by hand. Not only was that very expensive, but this dramatically limited the distribution. Today, you're used to going to one of many book stores and selecting a copy of one of tens of thousands of books, readily available at affordable prices—or downloading a book from the Internet without leaving home. Put yourself back in the mid-1450s. A common person had no books to look at, and didn't know how to read anyway. So the primary danger to the church from the Greek literature was from the wealthy elite few.

Gutenberg invented the movable-type printing press around 1455. [47] Of course, the first book Gutenberg printed was the Gutenberg Bible, but in the years to come, the technology of the printing press had an ever-increasing role in making books affordable and accessible—including, of course, the Greek literature.

So here we are in the Renaissance, the time of beautiful artwork, courtly manners and the birth of modern science. But the development of the arts and science affected only the elite few of the time, and the church and secular rulers were threatened by current events. How could the church regain its power? How could the rulers re-establish and maintain their control? Answer—the Inquisition. [48]

---

[47] *Johannes Gutenberg* entry retrieved July 25, 2006, from Encyclopædia Britannica 2006 Ultimate Reference Suite DVD.

[48] Professor Teofilo F. Ruiz. Professor at the University of California at Los Angeles. Terror of History: Mystics, Heretics, and Witches in the Western Tradition course http://www.thegreatcourses.com/courses/terror-of-history-mystics-heretics-and-witches-in-the-western-tradition.html

I know, it's not obvious. Professor Ruiz put it together rather well in his course called *Terror of History: Mystics, Heretics, and Witches in the Western Tradition*. For more details, I recommend watching the four-DVD six-hour video. You can check it out from your public library or purchase the set off the Internet. I will provide a much abbreviated version of the discussion here, with some amplification on salient points.

Why and how did Hitler target the Jews so successfully in Germany prior to WWII? What about the people targeted in the USA by McCarthy and his cohorts in the 1950s? Why did McCarthyism flourish? In fact, Hitler gained power by fanning the common hatred of the Jews due to their wealth relative to most Depression-stricken Germans. Adolf Hitler described Jews as "a parasite within the nation," and promoted the good of the Aryan people of Germany as more important than any other moral purpose. [49] Thus, he was able to re-distribute the wealth of the Jews in Germany, while providing a common enemy—the Jews. As you no doubt know, Hitler also used violence and fear to control those who might disagree.

Joseph McCarthy gained power by using the common fear of Communism. Anyone could be accused of being a Communist. Although not as dramatic or powerful as Hitler's actions in Germany, for a time, McCarthy was a man no one dared to oppose.

The Inquisition *Witch Craze* was similar, and probably more powerful than Hitler targeting the Jews. During the Inquisition, the church and dictators gained power by fanning the common fear of witchcraft. Remember that this

---

[49] *Adolf Hitler* entry retrieved July 25, 2006, from Encyclopædia Britannica 2006 Ultimate Reference Suite DVD.

fear was already common throughout Europe. You lived in constant fear that a witch might cause your son to become ill and die, or any other unfortunate event that might occur in your life.

The church to the rescue! Although the Inquisition was primarily concerned with true heresy, the church need only declare that witchcraft was heresy, so that the Inquisition could "save" the people from those dangerous witches. As the Jews would later become the common enemy in Hitler's Germany, so now, witches were the common enemy in Europe. Who were the witches? You could accuse anyone you didn't like of being a witch. Needless to say, you had to be careful not to make the wrong enemies, so it was best to accuse people who had little or no power—like older single women.

All right, so the church, through the Inquisition, could ferret out those dangerous old women, ah, witches, but what should be done with them? In 1486, *Malleus Maleficarum* (translated as Hammer of Witches) was written by Catholic clergyman Heinrich Kramer. This notorious book "was designed to aid them in the identification, prosecution, and dispatching of Witches." [50] The Inquisition used the *Malleus Maleficarum* to identify and prosecute.

Enter the secular authorities. They needed to demonstrate their absolute authority over life and death within their domain. The Inquisition would accuse and convict people, mostly older women, of being witches, and then turn them over to the local ruler for sentencing. Most of the time, the sentence was some form of execution, typically either hanging or burning,

---

[50] *The Malleus Maleficarum* (1486) by Catholic clergyman Heinrich Kramer. Transcribed by Wicasta Lovelace and Christie Rice. Can be viewed online at http://www.malleusmaleficarum.org/
http://bibliotecapleyades.lege.net/cienciareal/cienciareal12.htm

consistent with the *Malleus Maleficarum*. To emphasize their power, the rulers usually made attendance at these executions mandatory. In fact, these executions may have had something of a carnival atmosphere, as the people gathered from miles around to watch a vile creature being destroyed. [51]

Remember that the church condoned torture to gain confessions from heretics. Imagine your experience as you walk through a village where the Inquisition is currently active in the 16th century. You can hear the screams and moans from the old women being tortured, and you know some of the common means of torture. I won't be so crude as to describe them here, but various forms of torture and dismemberment are common practice in this era.

Now I want you to realign your thinking. You are not a common person. You are a scientist, attempting to discover the laws of nature in the 16th century. You have in your possession a copy of a Greek book. This is an incredibly valuable object, as it holds truth that would otherwise be unattainable. The common people around you do not understand—could not possibly understand—how important this knowledge is.

When you first gained access to this Greek book, there may have been no restrictions on your studies, but now the church seems to consider the topic of the book heretical. What should you do? This topic of study is perhaps the most important thing in your entire life. Well, last month, you heard that a friend of yours was killed as a heretic. You had been corresponding with your friend about this book, and he had shared with you his own related

---

[51] Professor Teofilo F. Ruiz. Professor at the University of California at Los Angeles. Terror of History: Mystics, Heretics, and Witches in the Western Tradition course http://www.thegreatcourses.com/courses/terror-of-history-mystics-heretics-and-witches-in-the-western-tradition.html

studies. Last week, you were forced to attend a similar execution, and you watched an old woman scream as fire devoured her body—after she had been convicted of heresy as a witch.

You have three choices. You can fight the church and continue to study the Greek knowledge openly—like your friend, whose name will not appear in any science book.

A second choice would be to publicly denounce the Greek knowledge, proclaiming loudly that it was a useless waste of time, but continue to study the book and its knowledge in carefully guarded secrecy. You can only hope to satisfy your intense curiosity, perhaps passing your learned knowledge onto the next generation. Your third option is to denounce the Greek knowledge, and find another topic of study, one of which the church approves. For myself, after denouncing the topic, I would continue my studies in absolute secrecy. What would you do?

Now bring your perspective back to the 21st century. Which scientists are to be found within the modern history and science books? Which of the three choices did they make? What type of bias does that give modern historians—and mainstream scientists?

I know that this part of our journey has been unpleasant, but I think it is important for you to think—and feel—as those early scientists probably thought and felt, so that you can better understand why they may have lied about the validity of the Greek knowledge of astrology, alchemy, magic and Hermeticism. Mind you, I'm not suggesting that all of these Greek writings were true. I'm simply pointing out that the scientists who made public statements about it, had no option but to claim that the Greek knowledge was false and useless.

But wouldn't the most clearly valid parts of the knowledge have survived in secret? The Inquisition was enforced with a death sentence off and on in Europe until 1834, about 600 years, [52] with the last auto-da-fé (the act of public penance of condemned heretics and apostates during the Inquisition) in Mexico in 1850. [53] Although the Inquisition was officially still in existence until 1908, heretics were no longer killed.

Magic and Hermeticism were important heresies as seen by the church. Although not as strongly targeted by the Inquisition, astrology was also an unsafe topic for scientists.

By the mid-1800s, only a few years after the end of the death sentence of the Inquisition, many Europeans showed open interest in occultism, especially Hermeticism. [54] Non-secret societies were formed, such as The Golden Dawn, today known as The Hermetic Order of the Golden Dawn (a well-known magical society). Secret societies, like the Rosicrucians, showed a public face again. Little is known of the inner workings of the Rosicrucians, for they have remained a secret society, but one of the high ranking members, William Westcott, gave sufficient hints in a recorded speech, "We must remember that Rosicrucianism itself was 'no new thing' but only a

---

[52] *Inquisition* entry retrieved July 32, 2006, from Encyclopædia Britannica 2006 Ultimate Reference Suite DVD.

[53] Professor Teofilo F. Ruiz. Professor at the University of California at Los Angeles. Terror of History: Mystics, Heretics, and Witches in the Western Tradition course http://www.thegreatcourses.com/courses/terror-of-history-mystics-heretics-and-witches-in-the-western-tradition.html

[54] Retrieved July 31, 2006 from http://www.hermeticgoldendawn.org

revival of still earlier forms of Initiation, and was a lineal descendant of the Philosophies of the Chaldean Magi, of the Egyptian priests, of the Neo-Platonists, of the Hermetists of Alexandria, of the Jewish Kabalists and of Christian Kabalists such as Raymond Lully and Pic de Mirandola."[55]

In spite of the revival of interest in astrology and magic, the direction of science did not change. Scientists had been forced to focus their attention on purely physical phenomena for hundreds of years, and they refused to acknowledge any new research related to nonphysical reality. What about modern times?

The Parapsychological Association was created at the suggestion of Dr. J. B. Rhine in 1957. Parapsychologists had been performing experiments and getting significant results for many years, and they continue to do so to this day. But no matter how many experiments are run with consistent results, mainstream science does not accept them.

Why don't mainstream scientists openly study such previously forbidden topics that hold verifiable validity? Let's look at that question in more detail on the next leg of our journey.

---

[55] Westcott, William Wynn. *The Rosicrucians: Past and Present, At Home and Abroad.* An Address to the Societas Rosicruciana. Retrieved August 1, 2006 from http://rosicrucians.ca/rosicrucians-past-present-home-abroad/

Section III: Control

Chapter 6: 20th Century

As we have seen, prior to the Inquisition, magic and astrology were part of science. During the Inquisition, a few scientists died after being convicted of heresy, but the founders of modern science consistently stated and taught publicly that any investigations of nonphysical reality were under the purview of the church—not science—and unworthy of study. Immediately after the Inquisition was abolished, many non-scientists again began to openly study other LORs from various angles, but scientists refused to consider any evidence that they had been bamboozled—just as Carl Sagan suggested.

My conclusion is that at least some of the founders of modern science lied to save their lives. You argue that there is a simpler explanation. You say, as you were taught, that it's all hocus pocus or imagination or delusion or slight-of-hand or such. Let's leave this debate open for now and consider more recent history. Remember that the scientific method requires us to balance skepticism and open-mindedness. Thus, we can't set aside either of these arguments until we gather more evidence.

Our journey continues early in the 20th century. My goal in this section is to determine, if we can, who or what is controlling modern science—or if you are correct, then there is no control involved. I have established a plausible reason why scientists in the Renaissance chose to denounce astrology, magic, alchemy and Hermeticism—to avoid torture and death at the hands of the Inquisition and the loosely allied secular authorities. Now that the Inquisition has ended, why do scientists still avoid these topics of study? Yes, Carl Sagan said that such a bamboozle is self-perpetuating, but is that the complete answer? Please bear with me, as we investigate this issue of control.

As I pointed out in the last section, the nature and direction of science as set during the Renaissance remained mostly unchanged into the 20th century. In 1905, Einstein's special theory of relativity shook the foundations of science, but they continued to look at purely physical explanations for phenomena. [56] Quantum physicists speculate about and are interested in nonphysical reality, but they are, as yet, not openly studying it. Why?

The Bible still demands that you kill any witch you come across, but the church has become more understanding if you don't take it seriously. The governments, at least those of the leading scientific nations, no longer execute people accused of witchcraft or heresy. So what are the roles of the church and state in modern science?

The numerous Christian denominations generally do not encourage scientific study of the previously forbidden areas of knowledge. Today, fundamentalist Christians are attempting to control science. Biblical creationists are still attempting to prevent, or compete with, the theory of evolution as taught in our schools. Some Christians still claim that the Earth was created in 4004 BC. But, honestly, these efforts seem to have very little effect on science. In fact, I have found no evidence that the church is controlling science to any significant degree.

---

[56] *History of Science* entry retrieved August 1, 2006, from Encyclopædia Britannica 2006 Ultimate Reference Suite DVD.

What about the government(s)? Are the governments of the high-tech countries made up of people, all of whom believe that nonphysical reality doesn't exist? Is that why they have such a united front, telling us to ignore such knowledge? No. That's not the answer. I'll show you what I think is a more realistic picture.

Our species is self-destructive. We fight and kill each other all the time. No, I'm not fighting anyone, and I'm sure you're not killing anyone at the moment, but our nation is usually involved in at least one deadly conflict or war, and there are always threats from people who would like to kill us. Our violence toward each other forces us to be on guard, and always be ready to fight—for survival, if not for aggression. All nations are constantly aware of this fact, the US more than most, and each government is always striving for two things: [1] as much information as it can get to determine when its people are threatened, and [2] superior weapons or related defensive capabilities.

In the US, the Central Intelligence Agency (CIA) is intended to be our primary source of information about threats from the world around us. Each country has its own army, and the US Army, not trusting the CIA, has its own intelligence gathering programs, as well as weapons development programs. These intelligence and weapons programs are mostly secret. To maintain that secrecy, the government usually denies that any such work is underway. In fact, the secrecy is generally on a need-to-know basis, so that the information is not available to government employees, even with Top Secret clearances.

No news so far, right? Pretty obvious to an educated person like you. OK, but did you know that the CIA and the US Army had long-standing secret programs to develop psychic abilities for both intelligence gathering and potentially as weapons? If you knew that, you probably thought it was just a legend or that they dropped those programs for lack of results. Naturally, the extent of these programs and their current status is not general knowledge. So is this just an unprovable legend? Do you think it falls into the same category as Big Foot, alien conspiracies, the Loch Ness Monster, etc.?

Think again. In 1995, the CIA declassified some of the research it had initiated in 1972 at the Stanford Research Institute (SRI), using a system called Remote Viewing (RV). [57] RV is a combination of techniques that facilitate more accurate psychic awareness under many circumstances. An RV session often results in descriptions and drawings of a physical location. (Remember my personal experiences from the first few chapters?)

In the early 1970's, the CIA apparently believed that the Russian military was experimenting with psychic techniques. Now it seems more likely that the Russians were working with physically based principles called "psychotronics" which use a combination of chemicals and electromagnetic radiation to influence people. [58] [59] This misinformation seems to have been the motivation of the CIA for initiating the research at SRI.

---

[57] *CIA Statement on Remote Viewing*. CIA Public Affairs Office, 6 September 1995. https://www.cia.gov/library/readingroom/docs/CIA-RDP96-00791R000100030062-7.pdf
https://www.cia.gov/library/readingroom/docs/CIA-RDP96-00791R000200180006-4.pdf

[58] *Statement by Ingo Swann on Remote Viewing* dated 1 December 1995. Retrieved August 8, 2006 http://www.ingoswann.com/statement-in-response-to-1995-cia-statement.html

[59] *The Russian View of Information War* article by Mr. Timothy L. Thomas. Foreign Military Studies Office, Fort Leavenworth, KS. Retrieved August 8, 2006, this

In 1995, the CIA announced that "the program" was "unpromising." Of course, this was after twenty years of research, and SRI had trained some of the intelligence community to continue using the techniques in secret. OK, that last part was supposition, and if this were all the public information we had to go on, I'd accept your criticism—but there's more. Before going into that, let me say a little more about the CIA-initiated RV research at SRI.

Although it is certainly interesting that the CIA admitted in 1995 to their participation in the Remote Viewing program at SRI, the results of those experiments are much more interesting. Yes, the SRI research was secret, but when the CIA declassified part of it, the researchers were free to publish those parts of their findings, and they did. Hal Puthoff was the director at SRI when the research began in the early 1970s. Russell Targ was another founder who participated in the remote viewing research. They have written numerous books on related topics. [60]

If you are interested, I recommend you read some of these books, but I wish to make a point here. I do not claim that their conclusions were valid, and by recommending them, I do not accept or deny any claims made in those books. These books by Puthoff and Targ provide valuable data for anyone

---

article was first published in *The Russian Armed Forces at the Dawn of the Millennium* (7-9 February 2000). Can be read online at http://fmso.leavenworth.army.mil/documents/Russianvuiw.htm

[60] *Mind Reach* by Russell Targ and Harold E. Puthoff (1997). *Miracles of Mind: Exploring Non-Local Consciousness and Spiritual Healing* by Russell Targ and Jane Katra, PhD (February 1998). *Mind at Large: Institute of Electrical and Electronics Engineers Symposia on the Nature of Extrasensory Perception (Studies in Consciousness)* by Charles T. Tart, Harold E. Puthoff and Russell Targ (November 2002). *Limitless Mind: A Guide to Remote Viewing and Transformation of Consciousness* by Russell Targ (January 2004). *Mind-Reach: Scientists Look at Psychic Abilities (Studies in Consciousness)* by Russell Targ and Harold E. Puthoff (February 2005).

interested in doing fundamental research. My point here is to remain skeptical *and* open-minded as required by the scientific method. (Probably most of their results are repeatable as they claim, but skepticism is an important part of scientific investigation.)

I mentioned the US Army, and additional public information—Major Ed Dames. [61]

Major Dames claims to have been one of the leading remote viewers in the corresponding US Army program, and he was decorated for some of his work as a remote viewer for the Army. Today, he is retired from the Army, and one of the premier civilian RV trainers. [62] You can learn more on his website. [63]

As you can see, the CIA and the US Army seem to have had programs in remote viewing, and today some of the participants are loudly proclaiming the success of those programs, while the US government remains quiet and unsupportive. The evidence also suggests the existence of other secret government programs using RV and other psychic techniques to both gather intelligence and use as a weapon.

Why would the government deny the existence of these programs and the degree of their successes? You probably already think as I do. Intelligence gathering techniques work best when the enemy doesn't know how you get

---

[61] Morehouse, David. (1998) *Psychic Warrior: The True Story of America's Foremost Psychic Spy and the Cover-Up of the CIA's Top-Secret Stargate Program*. New York, NY: St. Martin's Paperbacks.

[62] Who is Major Ed Dames? Executive Summary and Background. Retrieved August 8, 2006 from http://www.learnrv.com/eddames.cfm

[63] http://www.learnrv.com/

your information. And you're certainly not going to tell your enemies the details of mind control techniques you intend to use on those you capture. My point is that I do not criticize the US government for denying the existence of their secret intelligence and weapons research. Their silence is necessary, given the violent nature of the world we live in.

All right, so where are we? The church is no longer a significant threat to scientists. Although the government is not openly supporting private-sector research into nonphysical LORs, they have in the past, with great success. In short, the government is not preventing mainstream scientists from doing research into nonphysical reality. Clearly, the control changed some time in the last two hundred years.

But along the way, we discovered some additional evidence. Significant results were consistently achieved by SRI using RV techniques. Those results cannot be explained by accepted theories of modern science. This is a huge hole in your objection, and supports my theory. If you look a little further, you will find that significant results have also come from parapsychological research. Astrologers have been providing useful services to people throughout modern history and Carl Jung wrote many things about astrology, including, “I’m chiefly interested in the particular light the horoscope sheds on certain complications in the character. I must say that I have very often found that the astrological data elucidated certain points which I otherwise would have been unable to understand.”[64] In short, there is ample evidence from reliable sources to question modern scientific theories as they relate to other LORs.

---

[64] Roderick, Main. (1998) *Jung on Synchronicity and the Paranormal* (p. 11). Princeton, NJ: Princeton University Press.

So what prevents mainstream scientists from investigating nonphysical LORs and publishing their results? As I see it, a variety of factors influence people in general, and scientists in particular. Parents teach their children that only the physical LOR exists. They teach both by their actions and by their words, as I suggested when the small boy saw the energy from the power spot in Ohio, but then mimicked his parents and older brother in ignoring that phenomenon. School teachers continue this training, and reinforce it with science classes based on physical-only knowledge and beliefs. Historians believe what the founders of modern science wrote during the Inquisition concerning magic, astrology, etc., so history teachers reinforce this limitation. Scientists and historians support each other in their views.

To be fair, most people find it difficult to perceive even the etheric LOR. In general, perception is not straightforward, even on the physical LOR. For example, if you ask three witnesses of an accident to describe what happened, you will usually get conflicting stories. Thus, it is relatively easy to convince most people that any awareness they do have of nonphysical LORs is due to an overactive imagination or a temporary insanity.

These are generalities, and of course there are many exceptions. Some parents encourage their children to develop whatever awareness they may have. Some teachers encourage their students to "think outside the box." And some mainstream scientists are interested in investigating nonphysical LORs. So why don't they pursue their interest and publish their conclusions?

From an open-minded scientist's point of view, the control probably seems more immediate and not too subtle. If the scientist is a professor, he or she wants respect, prestige ... and tenure. The best way to attain these goals is to gain the respect of other scientists; that means solid research and publications. Thus, the scientist must publish in respected journals, but the people who control the content of those journals are not generally so open-minded, so the scientist must stick pretty close to traditional topics. This scientist is also restricted by funding, which is usually controlled by people who have the common traditional beliefs.

Let's consider the effects of simple indoctrination. Suppose you take a large number of children in several successive generations and tell them a consistent, plausible lie throughout every developmental stage. Those children grow up believing the lie and then teach the lie. The result is a type of social momentum. The children grow up to do the same thing to the next generation and so on.

Social pressure would be one implication of this type of social momentum. The children would be telling each other the same lie while they grew up, reinforcing it and supporting each other in their belief. As adults, they would still pressure each other to accept the lie. I also think if the lie were supported consistently by respected adults, any contradictory inputs would be ignored or at least the child would fight against them. I know there's a lot more to the psychology of this. Another implication would be what Carl Sagan said about a long-standing bamboozle—the resulting adults wouldn't want to consider any evidence showing that they had been misled and indicating that they were so credulous. We should look at it as a system. Once you get it going, it becomes self-perpetuating, like a large ball rolling down a long hill.

I found a related article on the Internet. It's an interview with Gregory Feist, author of *The Psychology of Science and the Origins of the Scientific Mind* and *Theories Of Personality*. Here's a quote from the interview. "Every field, on its path to maturity, goes through at least three stages. First is isolation, when a few founding figures or lonewolves are writing about an important new finding or perspective. These seminal figures are persuasive enough in their findings or arguments to attract a circle of followers, and then the field moves to its second stage, identification. The third stage of maturity happens when there is enough social momentum for graduate programs, societies, journals and regular conferences to form. This stage is called institutionalization." [65]

I think it's safe to say that these three stages of development were achieved for many fields of science prior to 1834—during the Inquisition. While Feist is talking about the development of fields of study, I suggest that the same is true for fields that are explicitly *not* studied; hence, I suggest that the choice by scientists to avoid studying nonphysical LORs was institutionalized during the Inquisition, with sufficient social momentum to perpetuate this restriction on scientific investigation into the 21st century.

My conclusion is that today everyone—and no one—controls the fields of study of mainstream science. The emperor's new clothes ... and our children keep telling us there's something wrong, but we've learned to ignore them.

At this stage of our journey, we can be pretty sure that we have constrained our science and technology. If so, we know how that constraint came to be and that we all participate in supporting the social momentum that perpetuates it. Even I share that responsibility by not publishing my findings sooner. Armed with this understanding, I contend that we can find a way to

---

[65] American Scientist Online (July 1, 2007). Greg Ross, *The Bookshelf Talks with Gregory Feist*. https://www.americanscientist.org/

move forward and expand scientific investigation so that we may control more aspects of our lives ... and maybe put some real clothes on the emperor.

In the next section, we will consider the obstacles to fundamental research of nonphysical LORs. The obstacles are many and difficult.

Section IV: Obstacles

Chapter 7: Obstacles to Fundamental Research

On this leg of our journey, we'll look at some of the reasons why scientific research into nonphysical LORs is difficult. As a starting point, let's consider the circumstances under which scientific investigation takes place.

The first step is to learn that something is there. I.e. as long as a person is unaware that nonphysical reality might exist, there is nothing to investigate. Most people are raised and taught to believe this, so it never occurs to them to seriously consider reports to the contrary and they are able to convince themselves that any psychic perceptions they may have are imaginary. I'm sure that many scientists believe that nothing nonphysical exists and that this is the most prevalent reason why few scientists attempt to study other LORs.

If a person suspects that the experiences are not just imaginary, he or she might decide to learn more ... but why? Why take the time and put in the effort? Others would think you're crazy (remember the social momentum thing) and what value would it have, anyway? Most people do not recognize sufficient reason to take action. What would it take to motivate a scientist to research and report on an area of nature where such reports might lower their professional prestige and jeopardize their career?

If you do think something is there, and you have a good enough reason for looking into it, how would you begin? Where would you learn more about it? How could you repeat relevant experience? Many books have been written on related topics such as magic, witchcraft, astrology, alchemy, psychism, astral travel, etc., but much that is in print is misleading, either because the author had misconceptions, which I think is most common, or because the author intended to restrict the communication of his or her knowledge to those who were able to see through the "blinds" [66] they wrote into the material. The misinformation makes it more difficult for the uninformed, especially for those who have no teacher.

The next step is perception. If you have no way to be aware of other LORs, even indirectly, you cannot observe the effects in those LORs of related experiments. Many researchers depend on related physical phenomena such as séances, Kirlian photography, or reports from psychics.

All right, suppose a person has sufficient reason to think something nonphysical exists, sufficient motivation to investigate, and some way to observe related phenomena, either directly or indirectly. What can prevent him or her from contributing knowledge of nonphysical LORs to mainstream science?

I'm sure there have always been a few scientists who have gotten to this point and investigated other LORs with varying degrees of success. In most cases, those scientists have not published their findings. Some of them may have studied with secret societies in which members swear an oath to die rather than disclosure "secret" knowledge. That would be more than sufficient reason not to publish—much as the Inquisition influenced the scientists between 1450 and 1850. As you can see, I think that secret societies contribute to maintaining the social momentum of mainstream science, but I cannot substantiate this claim. After all, I am not a member of a secret society, and if I were, I wouldn't be able to tell you about it.

---

[66] Fortune, Dion. (1972) *The Mystical Qabalah* (p. 1). London, UK: Ernest Benn Limited, Tenth Impression.

Some modern scientists have been free to publish their findings, and a few have done so. The psychologist C.G. Jung is one of the most well-known examples of this. Jung's groundbreaking work in psychology is well-known. He also studied paranormal phenomena and astrology, which he used in his practice. In a letter to Freud, dated June 12, 1911, he wrote: "My evenings are taken up largely with astrology. I make horoscopic calculations in order to find a clue to the core of psychological truth. Some remarkable things have turned up which will certainly appear incredible to you ... I dare say we shall one day discover in astrology a good deal of knowledge that has been intuitively projected into the heavens." [67]

Even though modern psychologists consider much of Jung's work in psychology to be important, Jung is not as well regarded because he had the gall to investigate and report on astrology and paranormal phenomena.

To be fair, we must remember that modern psychologists are the result of 400 years of Inquisition death threats followed by almost 200 years of social momentum and institutionalized reinforcement. Realistically, they would be crazy to think otherwise without good reason. OK, so maybe a lot of psychologists *are* crazy, but that's not the point. The point is that they are sufficiently in touch with reality to be influenced by other scientists who are affected by the long history of denial.

And that is the next obstacle to fundamental research into nonphysical LORs—the long history of ignoring related phenomena and experience in western civilization. A reasonable argument can be made that modern

---

[67] Letter to Freud (dated June 12, 1911) located at http://www.falconastrology.com/syn3.html

scientists have always concluded that only physical reality exists, so there is no reason to look at any related evidence. I'm sure many scientists used that argument in the 19th century shortly after the Inquisition was abolished, and it's still a common argument today. In the future, I expect we will use a more complete understanding of nature that includes other LORs, and we will develop technologies that dwarf what we have today.

The church is the next obstacle. The Catholic Church instituted the Inquisition to control Christian beliefs. Although this is not as much of a problem for today's scientists as it was 600 years ago, the church still resists scientific investigation into nonphysical LORs. Their approach is more subtle today. I don't claim to know all of their motives or methods, but I do see at least one form of church influence that is still effective. They promote the idea that psychic phenomena, and metaphysical knowledge in general, belongs to religion—not science. As mentioned earlier, mystical systems such as Wicca and the Qabalah often describe principles in religious terms, even though they are not fundamentally belief systems. Personally, I don't see why this argument of religious ownership is effective with scientists. If something exists, don't we want to learn about it? Isn't that the scientist's creed? What about all those movies about scientists who insist on gaining knowledge no matter what? Aren't those true? OK, maybe I've been brainwashed by TV. Nobody's perfect.

Governments present a different type of obstacle. The governments of technically advanced countries generally claim that science is justified in ignoring other LORs. The US government, for example, claims that such investigation is unwarranted and without value. This has major implications in funding and support for non-government research scientists. Most companies and wealthy people take their lead from government policies, so funding is seldom available from private sources, either.

With only the non-fatal opposition of peers, the church and governments, I think that scientists would still investigate and publish their findings of nonphysical LORs. I suggest that the next obstacle is really the first and most obvious—early training—the basis of the social momentum. Children imitate their parents from an early age. "Modeling a parent is one of the important ways a child learns social roles and the behaviors and skills judged appropriate to those roles." [68] Parents teach their children that perception of anything nonphysical is wrong and inappropriate. "Johnny has an imaginary playmate, but he'll grow out of it." In most families, awareness is not encouraged—the children are raised to be "normal" and the norm was set during the Inquisition when too much awareness could be fatal. Early training gives a person a sense of what is right and wrong that lasts throughout his or her life, and it's considered "wrong" to insist that you see things that others do not. I contend that this adds to the pressure to conform, which reinforces the social momentum of mainstream science.

"Teachers are by far the most important influences on schoolchildren." [69] Teachers reinforce these beliefs and associated behavior throughout primary and secondary school, and continue it in college. Teachers tell the child that he or she is illogical (i.e. wrong) to believe in things that are "not real," and give scientific opinions to support their point of view. Students generally

---

[68] Hetherington, E. Mavis. Parke, Ross D. and Locke, Virginia Otis. (2002) *Child Psychology: A Contemporary Viewpoint* (p. 608). Boston, MA: McGraw-Hill College.
[69] Ibid.

trust their teachers and accept what they are taught as fact. Since they've already been taught that their own perceptions are wrong when they disagree with the norm, most of the children are easy to convince. Those who continue to perceive nonphysical phenomena learn to keep quiet about it.

As you can see, I'm suggesting that scientists are human, having been influenced (mayhap brainwashed) by their parents and their teachers before they became scientists. Thus, when their peers do not support their interest in unsponsored research, it is difficult for a scientist to fight against their peers and their training, and actually publish the results.

Which brings us to another obstacle—a big one. If research is not published in one of the respected and acknowledged journals, it is mostly ignored by mainstream scientists, no matter how important the report might be. And if the report contradicts the beliefs of most of the screeners for a journal, the results aren't published there. This is part of the social momentum, the institutionalization of science. Most of the time, this type of exclusion is good because it reduces the irrelevant information, or even disinformation, that mainstream scientists do not wish to waste their time reading. Unfortunately, as regards nonphysical LORs, the disinformation is being perpetuated by the very system that was designed to eliminate it.

But even with all these obstacles, if scientists saw consistent, interesting and verifiable results, I think they would push forward with the research, breaking down the barriers to publication, and drawing nonphysical LORs into the "system" that we call nature. I contend that scientists find the research to be quite difficult, and that the difficulties of the investigation support the social momentum to ignore other LORs.

Charlatans. What part do these insincere people play in scientific research? They muddy the water. If a scientist cannot make direct observations of other LORs, he or she must depend on others. Charlatans often make it seem that dramatic events are real, when in fact, they are slight-of-hand tricks. For example, charlatans often produce phenomena at séances. I'm not claiming that séances cannot have real phenomena, but that is my point. How is a scientist to tell the difference between what is real and what is faked? The claims are the same whether the "medium" is real or a fake. If no wires are found or fog machines or other physical bases for the observed phenomena, the researcher can't tell the difference. Then if the researcher publishes the finding and another scientist figures out how the first one was tricked, all credibility is lost. In short, charlatans support the social momentum by confusing the issues.

I suspect that if causal factors were clear and well-understood, charlatans would not be able to ply their misdirection. Unfortunately, without basic research to discover the nature of the LORs, we can only guess at the causal factors behind the phenomena we observe. This would seem to be a catch-22.

Lacking both scientific support and an understanding of causal factors, experimental procedures are difficult to define. For example, how much does the attitude of the researcher affect the results? Some experiments have shown that the experimenter's expectations do affect the outcome. How? Why? To what extent? Can an unconscious attitude affect the results? How can an experiment be designed to exclude such perturbations? This is only one example of the complications in designing such experiments.

Even with all these difficulties, the greatest obstacle to the scientific investigation of nonphysical LORs is probably the inability of researchers to make direct observations. If everyone could see auras (an etheric-level aspect of plants and animals), no one would deny their reality, and experiments would be easy to design and run. Scientists have worked successfully with other invisible phenomena, such as molecules and viruses, but in those cases, physical equipment has been created to make mechanically assisted observations. Kirlian photography may be such a mechanism to provide assisted observations of auras. If so, that would provide a starting point, but without the ability to observe more than auras, the research would still be quite limited. Thus, lacking the ability to directly perceive characteristics of other LORs, scientists find it easy to deny their existence, which brings us back to the first obstacle—as long as the scientist believes nothing is there, there is nothing to investigate.

At this point in our journey, you have seen ample evidence that other LORs do exist. You have seen why those LORs are not investigated by today's institutionalized mainstream science. The obstacles are many and daunting, like a series of brick walls, and you have only your head to break them down. You have also seen that the other LORs can and do affect our lives. I contend that somehow, some day, scientists will break the cycle and expand our knowledge of nature and our ability to control our lives. That day may be today. Let's break through those brick walls together.

Chapter 8: Overcoming the Obstacles

We've seen that modern scientists have many good reasons to ignore and/or doubt the validity of any evidence of nonphysical reality. The founders of modern science claimed that all related knowledge was false and useless, and no respected scientist has shown proof to the contrary. Well, OK, there have been a few well-respected scientists such as Jung who have presented strong evidence, but we might dismiss them by calling them eccentric. Among scientists, no one can see or otherwise perceive anything nonphysical. Well, OK, there are a few exceptions to that rule, too, such as Puthoff and Targ. The effect sizes in experiments have been very small, with most parapsychological experiments requiring thousands of trials to get significant results. And, yes, there are counterexamples to that rule, as well. But you see the point. Most modern scientists have grown up being taught from childhood that nonphysical reality simply does not exist outside religions, and they are taught *not* to perceive anything nonphysical. From their point of view, this is perfectly rational and reasonable, and the social momentum supports them.

This seems like a Gordian knot. How can we possibly break a social cycle like this one? It's been going on for nearly 600 years.

First, I think it's important that a researcher develop the ability to make direct observations, as I have. In most scientific fields, such as physics and chemistry, the researchers make direct observations whenever possible. No one came to Newton and told him about an apple falling on their head. An apple fell on *Newton's* head (or he watched one fall)—a direct observation.

Does a chemist depend solely on others to tell him or her the color of a chemical solution? No, of course not.

Equipment often provides enhancements to the researcher's perceptions. An electron microscope, a chromatograph, a voltage meter—these provide reliable extensions to the researcher's physical senses, but the researcher uses them directly. He or she does not depend solely on other people to take the readings. Even when assistants are employed, the researcher can still take some direct readings to spot-check.

Even in psychology, the researcher makes direct observations. Since psychology is currently the study of human thought and behavior, the researcher observes human subjects. If the subjects know what the researcher is attempting to discover, they may bias the outcome, but psychologists know that and are able to mislead or misdirect subjects so as to obtain less biased results.

If the researcher is investigating other LORs without being able to perceive those LORs, the results depend heavily on the ability and reliability of the people actually making the observations or creating the observable phenomena. If such a person is a charlatan, the results will be biased by the agenda of the charlatan. Even when the motives of the assistants are pure, the results can be biased by their desires and expectations.

The generation of ideas is another consideration. Newton thought about gravity because he made a direct observation, and that thought led to his theories and subsequent experiments. I am suggesting that many theories derive from unexpected events or side-effects. In the process of direct observation, the researcher may notice something that another person would not consider important or interesting, leading to a new or different theory, with an interesting and testable hypothesis, whereas an assistant might fail to relay the needed information.

These are some of the reasons why I contend that each researcher should learn to directly observe nonphysical LORs and never depend solely on the perceptions of other people. I have mentioned three systems to increase ones awareness of other LORs. These are Silva, Remote Viewing, and my system, which I call AWIN. I will discuss RV and AWIN in more detail in later chapters. In fact, there are other systems for the development of etheric vision and astral travel, of which AWIN is an example. I consider Silva to be a more limited and potentially limiting approach. The Silva system includes some techniques that could prevent you from making progress beyond a certain point, as I personally witnessed in one student. However, the Silva system is effective for medical-related research, using the Silva "cases" method. Thus, I only recommend it for this specific type of research.

Between RV and effective etheric vision/astral travel systems (e.g. the AWIN system), each approach has its own advantages, and both are effective for making direct observations in scientific experiments on the etheric and astral LORs. Since RV is supported by books, web sites and several organizations, I will only devote one chapter to it. I do not claim significant familiarity with other systems for the development of etheric vision and astral travel, so I will focus on the AWIN system. In the absence of a network of qualified instructors, I will describe the AWIN system in some detail. However, I will only describe the portions of the AWIN system that are needed for the development of etheric vision and astral travel. Those two features are of immediate relevance to fundamental research in nonphysical

LORs, while the more advanced features of the AWIN system would only be distracting at this time.

Most people can develop a functional level of awareness in about 30 days with either RV [70] or the AWIN system. Researchers willing to put in one month of effort will be able to begin to make direct observations on the etheric LOR. Those same researchers spent many years of concentrated effort in college, learning about their fields of endeavor. This additional month is insignificant in comparison. With this small amount of time and energy, a scientist can overcome the most significant impediment to fundamental research in nonphysical LORs. Naturally, this is only a starting point, and those interested in broader research into other LORs will need to make periodic effort to maintain and increase their ability to perceive, first the etheric LOR and then the astral LOR.

You may ask, "Why didn't people know about these systems before?" Actually, RV has been widely used for at least twenty years. The AWIN system has only been known to a few students, but it is based on the Qabalah, which has been in use since before the Inquisition. Mostly what I have added to form the AWIN system is a combination of a scientific approach and a lack of secrecy. While I have added a few more advanced techniques, the Qabalah contributes most of the principles and techniques used in the AWIN system. So you see, systems for learning to perceive other LORs have been around for a long time, but mainstream scientists didn't know those systems had any validity because they believed that other LORs didn't exist.

---

[70] Learn Remote Viewing FAQ retrieved December 30, 2006 http://www.learnrv.com/faqs.cfm/

Which brings us to the most basic problem in this Gordian knot. Although I think I've established that nonphysical LORs do exist and are important in our lives, most people, especially scientists, may contend that all I've done so far is compile a plausible justification with references to research that they don't believe, anyway. If you are one of those people, I suggest that you skip ahead to the first astrology experiment, and run it. Astrology is the study of the correlation between the positions of heavenly bodies and influences on our lives. A basic characteristic of astrology is the position of the sun at the time of birth as it influences the personality of an individual—throughout his or her life. If this correlation is valid, it suggests the existence of other correlations, which, in fact, is what astrology is all about.

However, if you, like C. G. Jung, do not see the relationship between astrology and other LORs, I understand and suggest alternative ways for you to prove to yourself that nonphysical LORs exist. Using the AWIN system, a person can perceive and manipulate astrological influences as I have. Without that link/experience, I realize that astrological correlations appear consistent with Jung's theory of synchronicity [71] with no apparent causal factors.

If you do not wish to learn to make direct observations, and you are not intrigued by the definitive results of the astrological experiment, existing research results are really the best source of proof. I suggest you read some of Dr. Dean Radin's work. For example, try *The Conscious Universe: The scientific Truth of Psycnic Phenomena*. Dr. Radin has also published many journal articles on his research. He is one of many strong scientists who

---

[71] Roderick, Main. (1998) *Jung on Synchronicity and the Paranormal*. Princeton, NJ Princeton University Press.

research the "paranormal." Based on the historical perspective and the associated social momentum discussed herein, now you understand why their work has not been incorporated into mainstream science.

All right, I'll assume at this point that you are convinced of the existence of nonphysical LORs, or at least you concede that they probably exist. Now, why should you care? If you're like me, you're curious and you don't like receiving (or giving) misinformation. It's important that you get involved, and you may not feel the same way I do, so here's my pitch. I have three more reasons for you to get involved. First, many people will have to contribute to break the cycle, so your contribution is important.

Second, this is important to your life and to the lives of those you love. Sound fishy? Here are a few benefits to consider. If the simple astrology experiment shows the direct correlation that I predict between a person's sun-sign and personality, this has major implications for any couple considering getting pregnant, meaning that they might prefer to live with a child whose personality is compatible with their own.

If that correlation exists, then some other astrological correlations are probably real as well. For example, I lived with a woman when she had a major transit that brought violence. It's a long story, so I'll make it short. She had a car accident wherein our car was totaled. Most strangers were reacting to her with animosity. She was almost arrested for hit-and-run when she scratched a neighbor's car as she got out of her rental car and proceeded into our apartment. That's when I looked at her transits and found the problem. With my help, she eliminated the influence in a matter of minutes, and had no further effects from that transit. Experienced astrologers consider that particular transit to be fatal a significant percentage of the time. Wouldn't it be nice if you could talk to a certified specialist to help predict and solve such problems in your life? Maybe learn enough about things that influence us so that you can have more control over your life?

At this point, you may be thinking that astrology is all you need to know about, but it's not that simple. Strong astrological influences can be observed in a person's aura, but other things affect a person's aura as well, such as other people. Have you ever known anyone with whom you had a strong personality conflict? Did you notice that you could tell when they walked into the same room with you, even when you didn't see or hear them? This is a simple example of a broad category of influences.

Psychics also say they can often see disease in the aura before it becomes serious. Wouldn't it be nice if certain doctors were certified to observe such changes in your aura? You would be healthier and live longer. OK, but you still may get sick. Wouldn't it be nice if certain doctors were certified to use a version of the Silva cases, so they could diagnose problems without invasive surgery or expensive tests?

So far, I've focused mostly on some of the potential benefits to your physical health and longevity. Health is important, and that includes mental health. From personal experience, I know that it is possible to directly access another person's subconscious, using the AWIN system techniques. With these techniques, a qualified psychologist would be able to help you solve long-standing problems of many kinds, e.g. in relationships or repeated negative patterns in your life. We all have such problems, and if they are severe enough to consult with a psychiatrist or psychologist, you probably want the problems to go away. Today, schools don't teach psychology courses with techniques providing direct access to a person's subconscious. They don't believe that such things are possible.

In my view, achieving major health improvements is a good enough reason to want these technologies to be incorporated into and studied by mainstream science. But quality of life is another important area. Without health, there is no quality of life, but even with excellent health, what do you need to do to be happy? By understanding yourself better, you can learn more about what you need, and how to get it. I.e. if you know more about what and who you are, you are better able to make decisions and take actions that increase your happiness. Wouldn't you prefer to look back at your life in twenty years and say, "I wouldn't change a thing."

But maybe you're a technologist or a business person. You have good health, and you don't think you need any assistance in your life. In fact, you don't think anyone you love needs it, either. What you're looking for are technological applications as a basis for a thriving business. Personally, I haven't looked much into these types of applications of nonphysical LOR knowledge. I have cared primarily about health and well-being, for myself and for everyone—but I understand your perspective, and I'll relate some suggestive principles and experiences that may be helpful.

In the computer field, processors are limited in their processing power by the speed of light. As I indicated earlier, the speed of light does not seem to be a limitation on the etheric and astral LORs. If that is true, and you were able to build a stable processor on the etheric LOR, it might be thousands or millions of times faster than current computers. Of course, it would only be useful if the data and results could be communicated between physical equipment and the etheric-level processor.

In my experience, a microwave emitter influences the etheric level. Kirlian photography is also claimed to record etheric-level objects, like auras. If these claims are true, it means that at least some methods do exist to create non-animate physical machines that can interact with the etheric LOR. What if I am correct, and both mass and energy can be translated between LORs? The implications may be huge! We have an ever-increasing need for energy on this planet, and at the same time, the atmosphere is converting sunlight to heat, causing global warming. Pardon me while my imagination goes crazy on this! What if we could shift heat energy in one place to the etheric and then use it to run a machine, maybe an electric generator, in another place? Or maybe we could shift sunlight to the etheric before it reaches the Earth's atmosphere. Energy could be supplied to anywhere in the world without wires or any other physical medium. The power companies would love it.

Consider the applications of shifting matter between LORs. Transportation. We would be able to travel all over the world without airplanes. The cost of shipping goods would be ... all right, if we can build such machines, we don't know how much they would cost or what they would need to function, so we can't guess the cost of transportation using them. Oh, and what about interstellar travel? Our astronomers and other scientists could travel to other star systems, other galaxies, and study phenomena first-hand. And our resources. We could dump our garbage into the sun and get fresh resources from other planets. I *like* it! You can imagine many more applications for interstellar travel, so I'll stop with those few ideas.

I'm sure there are lots of other ways we would benefit by incorporating knowledge of nonphysical LORs into our science and technology, but I think I've given enough reasons to motivate most people to begin the investigation.

The third reason for you to get involved is personal gain. Many people will pioneer specific interesting developments as we investigate nonphysical LORs. Those people will benefit in many ways. You will see books published, research grants given, faces in the news, patents granted, etc. Think what a patent might be worth to you? What if you were the one who discovered and patented a method to create an etheric-LOR computer? Better yet, what if you discovered a means to shift physical energy to the etheric and back to the physical? Actually, the possibilities are endless, and many people will benefit from their efforts to discover the nature of other LORs.

Let's say that at this point, you know that nonphysical LORs probably exist, you have sufficient motivation to care, and you know how you can learn to make direct observations in scientific experiments. I've given enough references to provide useful reading to learn more about related phenomena and principles. What about the other obstacles? First, none of the information or suggestions I've given here are secret, so you are in no danger if you move forward. Well, if the church reinstates the Inquisition, you may need to back off, but that's unlikely today. As to peer-pressure, we *will* be criticized initially for investigating these long-forbidden topics, but as more of us actually look and acknowledge the validity and usefulness of other LORs, the criticisms will fade and be replaced by eagerness as more scientists begin to find answers to questions that have no answers in the physical LOR.

When they realize that they can no longer hide its validity behind secrecy, the governments will quickly back civilian research. I'm sure they will claim that they didn't know, or make other face-saving comments, but the result will be the same—government funding will become available, as all countries scramble to take the lead in the new technologies.

The church may be the last to accept non-religious investigation of other LORs. Many psychics make religious claims about what they find on other LORs, and that may make the churches nervous. After all, if there really is a truth to be found, our scientists may actually find it. Then the churches would need to change to keep their followers. Personally, I prefer reality—whatever that happens to be. I'm not saying that the scientists *will* find the truth, but I suspect that it's there to be found, and given enough time and enough researchers, I think science can uncover it.

Next obstacle—beliefs learned as children and taught in schools. This will take time to overcome, but as mainstream scientists begin to learn about other LORs and technologists begin to develop practical devices based on that knowledge, parents and teachers will quickly change their attitudes. How long did it take for the Internet to catch on and become an inherent part of our society? I suggest that a similar transition will take place.

To publish your experimental results, we may wish to create a few journals, some for standard scientific research and others for supportive research. The new standard science journals must uphold high standards for the research and the reporting. The supportive journals should be available for those who can contribute new findings or new information, but may not stick with the highest standards of the scientific method, or know how to write up their results to meet the higher standards. I think we need journals to report all levels of research, not limiting the research to people with PhDs.

As physicists and psychologists begin to find answers to some of their unanswered questions, they will also publish in the current mainstream science journals, but I expect our new knowledge to be radically different from existing fields of physical studies. By publishing in these new journals, we can rapidly institutionalize mainstream scientific investigation of other LORs. Soon thereafter, we will need journals for those developing related technologies. By creating new science and technology journals, we can also bypass those who screen articles, continuing to support the slowing social momentum by thwarting publication of the new research.

Charlatans will no longer be significant obstacles. When researchers are making their own direct observations, charlatans will be recognized and ignored.

Lastly, we come to the issue of experimental procedures. I contend that through the use of direct observation and the replication of experiments, many of the subtler problems will become less important. I am not suggesting that procedures should be lax. In fact, I support the use of stringent controls. I simply suggest that when the controls are not practical, useful experiments can still be performed by careful and thoughtful people.

I would like to pause for a moment and consider some implications of our journey at this point. At the end of the Middle Ages, physical phenomena were not understood, thought of largely as being unknown and unstable, potentially modified by the whims of Gods or magicians or witches. Scientists began to investigate physical phenomena and found regularities, consistent principles such as gravity, mass, electricity, inertia/momentum, chemical compounds, chemical reactions, etc. These were principles that had always been there, and simply needed to be uncovered by enquiring minds. In short, science turned the unknown physical LOR into what we have today—a stable and largely controllable environment.

We have seen that our current view of nonphysical LORs is similar to the view people held of the physical LOR less than 1000 years ago. There are differences, of course, but do you see the similarities? And now that the obstacles have been removed, the nonphysical LORs lie open for our investigation—a scientific opportunity that defies estimation.

By way of a starting point, several later chapters describe a wide variety of experiments, including suggested criteria for designing additional experiments. But first, let's look at the nature of science—what does it do for us today, and how might our lives change with the coming research into other LORs?

Section V: Science Today and Tomorrow

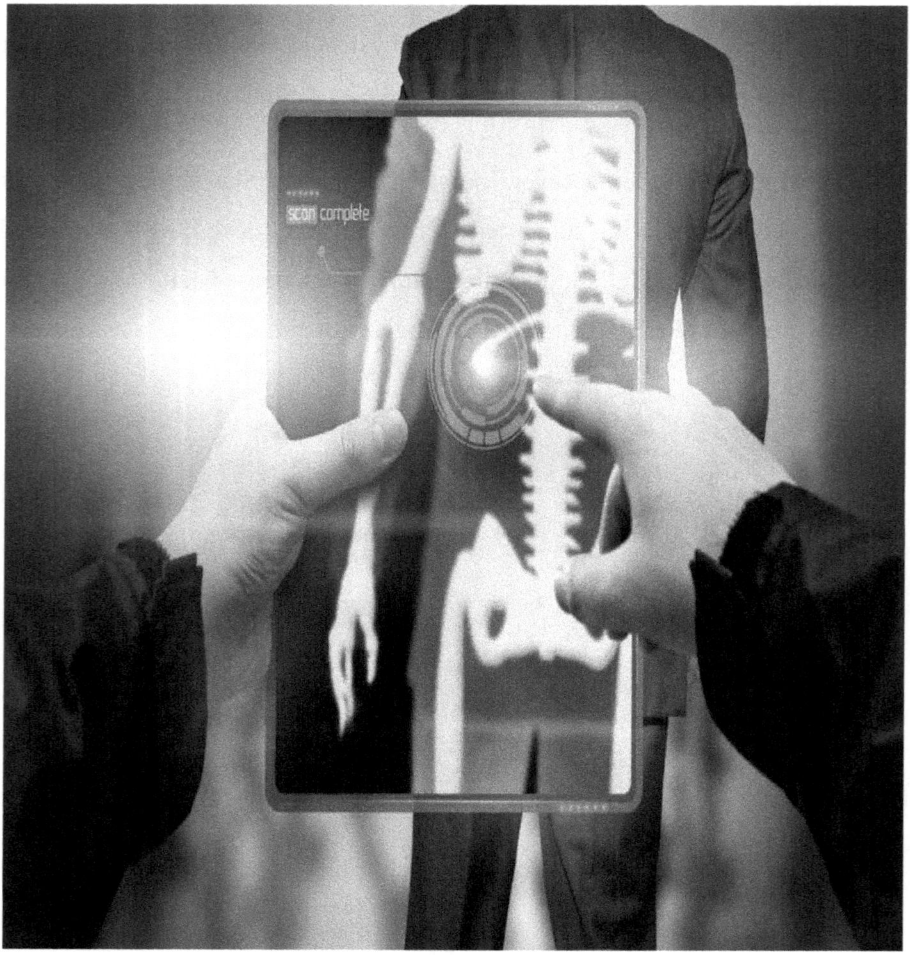

Chapter 9: Science Today

On this leg of our journey, let's take a brief look at science and technology as they exist today. The advances in our knowledge of the physical LOR in the last fifty years have been tremendous. The value to humankind is undeniable—but so are the remaining limitations.

Computer technology is the most obvious. Computers are based on something called a transistor, which has certain electrical properties. When computers were first conceived, they were thought of simply in terms of calculating numbers—arithmetic. In fact, computers were used for that limited purpose for some time before researchers realized that they could do more than arithmetic. Yes, believe it or not, word-processing and spread-sheets were not available on the first computers. Neither was digital photography or music. Letters were created on something called a typewriter, which you may never have seen, and probably have not used. Today, communication takes place mostly via cell phones, computers and the Internet. People used to write a letter, place it in the mailbox by the curb and expect it to arrive at its destination in a day or so. Today, you write an email and expect the recipient to get it within seconds. People used to go to the library to research a topic of interest, such as for a school project. A small research project took at least a few hours. Today, if you want to know something, you look it up on the Internet—in a few minutes.

Computers are used to automate and control many devices. An automobile typically has several computers, controlling such things as fuel-air mixture, injection rate, braking, navigation and more. Needless to say, many computer peripherals have embedded small computers, such as scanners, printers, routers, etc. TVs and stereos have hidden computers. Clocks, microwave ovens and many other household appliances are likely to have computers that control their operation.

Industrial automation has changed many aspects of our society. As workers, we hate when our jobs are automated. Our automotive industry is a prime example of this phenomenon. However, as consumers, we love automation. By automating industrial processes, we have achieved higher quality and greater consistency in our products. Only thirty years ago, some people selected a car based on the day of the week on which it was manufactured. Workers were thought to do a better job on certain days, and not so good on other days. At the time, this may have been an erroneous philosophy for selecting a car, but it hardly has any relevance today, since workers have much less effect on the cars being manufactured now. Highly specialized robots do many of the production and assembly jobs in factories, and the number of robots continues to increase. Again, as workers, this is a bad thing, but as consumers, we love it.

As mentioned before, computer technology is limited by the speed of light. Electrical impulses travel at close to the speed of light. You may have noticed that the speed of computers used to double every couple of years. The increases in processing speed were largely due to greater miniaturization. By making the paths between components shorter, their interactions take less time. As our computers became more powerful, we found ways to use them that required even more speed. But today, our computer circuits are almost as small as we know how to make them. The result is that with our current computer technology, we have been unable to increase the speeds significantly for several years. Computer games, graphical processing, databases, and many other computer applications have continued to evolve, demanding more and more speed—that will not be forthcoming with the current technology. A radically different technology will be required.

Scientists have been working on these problems for many years. Molecular-scale processors would give us a leap in performance, but that's as small as we can get. Even if that technology is practical, the speed of light would still limit the processing power of computers.

Other technologies have spun off from computers. The Star Trek communicators, for example, can be seen everywhere—cell phones. When a person next to you on the street speaks, it's probably not to you. The teenager is gossiping with a friend who is on the other side of the city. The business woman is arranging a meeting or strategizing with a coworker while on the way to the office. In short, cell phones have increased our availability as well as the rate at which we communicate.

Years ago, if you wanted to see a movie, you went to a movie theater and watched it with a crowd. Today we not only have VCRs (remember those?), we also have Tivo, high-def, DVD-recorders, movies-on-demand, and more. Radio has changed. Now we download hundreds of songs on our portable players and listen to the music we prefer, when we want it.

A businessperson used to have a personal planner that he or she carried everywhere. Now that book is replaced by a "BlackBerry." This is a portable device that combines several highly useful features, such as email, instant messaging, web browsing, and more. With such a device, a person can go to meetings or travel, and keep up with communication without a hitch. Today, you can get a BlackBerry device that fits in your shirt pocket and includes personal data assistant (PDA) features and is also a cell phone. It's like having a personal secretary with you at all times—and the library of Congress and direct access to everyone you may need to speak with.

Our genetic research and technological advances are approaching computers in their level of importance. Today we are able to perform DNA analysis on a small amount of biological material, and match that biological material to its source. Thus, we can determine not only what type of plant or animal it came from, but even which specific person. This has important applications in law enforcement, for example.

Genetic research includes gene cloning, recombinant DNA technology, gene therapy, and the genetic engineering of bacteria, plants and animals. The Human Genome Project is a scientific research effort to analyze the DNA of humans and of several lower organisms. It began in the United States in 1990, and in 2000, participants announced the completion of the initial sequencing of the human genome, which is composed of about three billion nucleotide base pairs. [72] Based on this data, researchers are better equipped to understand the functioning of the human body. What causes bones, muscles and organs to grow? What are the nature and cause of genetic diseases? Why is one person able to fight off a disease while another person dies? These are some of the questions that are more easily answered based on an understanding of the structure of the human genome.

---

[72] *Human Genome Project* entry retrieved July 6, 2006, from Encyclopædia Britannica 2006 Ultimate Reference Suite DVD.

Perhaps more importantly, scientists are attempting to use the data from the Human Genome Project to plan how to modify human genetics. Small changes in a woman's DNA might make her less susceptible to certain classes of diseases, or more capable of carrying a baby to term. A man's bones might grow stronger or his muscles might move faster. In fact, the possibilities are immense—for understanding and for change. Could a person grow gills to swim underwater for long periods? What about eyes like an eagle? Or maybe the ability to hibernate for months at a time? These are only a few of the potential human genetic research directions.

What about other genetic research? How about new types of plants? Maybe a plant that would grow in most climates, absorbs greenhouse gases from the air and whose seeds are high in protein. What about animals? We can clone animals. We can modify their genetic makeup.

Obviously, I can make no attempt here at completeness when describing our science and technology, and my description of genetic research is a prime example of brevity. Neither will I attempt to cover the ethical and safety issues involved in genetic research, but I think a couple of examples should be indicative. Suppose researchers come up with a new species of animal. In laboratory conditions, it is docile and easy to handle. It will eat most types of vegetation and grows rapidly, with little fat content. Sounds great! This type of animal could be raised in many areas, providing low-cost protein to the human race. But we could not know what impact the new species might have on the ecosystem of a given area, nor could we know how it might change when introduced into various habitats. It might begin to eat other species. It might attack and kill people. It might become so prolific that vegetation would become scarce. It might spread disease. It might evolve rapidly outside the laboratory into—what? We don't know. We can't even be sure of the long-term effect ingesting its flesh would have on humans.

So how about a simpler organism? In 1981, Ananda M. Chakrabarty patented a bacterium that breaks down crude oil. [73] Today, he is working on a bacteria-based cancer drug. [74] Obviously, these are good things—but. Bacteria and other micro-organisms tend to evolve more rapidly than larger animals. Genetically engineered bacteria might evolve in nature into something deadly or otherwise destructive. Probably not, but we just don't understand enough to be confident.

We don't have all the answers. We have only just begun to learn about the tremendous power and complexity of genetics. The potential impact of genetic-based technology is explored to some small degree by science fiction novels such as *Next* by Michael Crichton. Our imaginations go wild when we consider what might be done with DNA. Genetic engineering represents great power. Like atomic power, it could be used to build better lives—or to destroy the world we know.

As always, astronomy is an important area of our science. It has been a key aspect of science throughout history. People have always wanted to know the origin, past and future of the Universe, for example. Astrophysicists have studied the emissions and movement of the stars with ever-increasing

---

[73] Ananda Chakrabarty, MIT web site information retrieved January 4, 2007
http://lemelson.mit.edu/resources/ananda-chakrabarty
http://scienceblogs.com/oscillator/2010/06/08/oil-eating-bacteria/

[74] Chakrabarty, Ananda M, Nuno Bernardes, and Arsenio M Fialho. *Bacterial Proteins and Peptides in Cancer Therapy: Today and Tomorrow*. Bioengineered 5.4 (2014): 234–242. Web. 13 Aug. 2017.
https://www.ncbi.nlm.nih.gov/pmc/articles/PMC4140868/

accuracy. They predicted the origin of the Universe, hypothesizing that all physical matter came from a single super-dense lump that exploded long ago. Based on that hypothesis, they predicted a particular type and amount of radiation, which was then verified, supporting the theory now known as the Big Bang.

Astrophysicists have found evidence of many principles of the formation and lives of stars, planets, and black holes. Based on their theories and evidence, they estimate that the Universe began with the Big Bang between 10 and 20 billion years ago, [75] most likely close to 15 billion. According to NASA, "... the age of the Earth and the other solar system planets is about 4.5 to 4.6 billion years." [76] Our Sun is estimated to be about 4.57 billion years old, [77] [78] about half-way through its life cycle. [79] We all want to know that our sun will continue to shine for a long time to come, so this is good news.

We also want to know if we are alone in the Universe. Many people have reported seeing UFOs that may indicate the presence of extraterrestrials near the Earth. If there are aliens here, they're not broadcasting on public radio or giving presentations at United Nations meetings. Maybe the government really is hiding a pact with aliens, but whatever the situation may be, as a

---

[75] Big Bang Theory: Evolution of Our Universe article retrieved January 1, 2007 located at http://imagine.gsfc.nasa.gov/docs/science/ mysteries_l1/age.html
[76] retrieved January 1, 2007 from http://solarsystem nasa.gov/faq/index.cfm?Category=SolarSys
[77] Bonanno, A., Schlattl, H. and Paternò, L. *The age of the Sun and the relativistic corrections in the EOS*. Astronomy and Astrophysics (2002) 390: 1115-1118. https://www.aanda.org/articles/aa/abs/2002/30/aa2598/aa2598.html
[78] What Is The Life Cycle of the Sun? article retrieved January 1, 2007 from https://www.universetoday.com/18847/life-of-the-sun/
[79] Pogge, Richard W. (1997) *The Once & Future Sun* (lecture notes). Retrieved January 1, 2007 from http://www-astronomy.mps.ohio-state.edu/~pogge/Lectures/vistas97.html

race, we're not sure. The church tells us that we're unique and that no other life exists on as high a scale as humans. What is the truth? We would all like to know.

The human population has become an ever-increasing problem on this small planet. Our technologies, such as computers and automation, have become energy hungry. We want our living spaces cooled when it's hot and warmed when it's cold. We want to live at a distance from our places of work, requiring energy-hungry transportation. We are using up the natural resources, such as metals, oil, etc. Our chemical processes result in pollution that enters the environment. Our appetite has dramatically cut the number of fish in the ocean and changed the ecological balances. If we numbered fewer, our planet could support us easily, but we continue to procreate faster than we die. What should we do about this problem?

If we could travel to other star systems, we might be able to spread our population across the galaxy, much as on Star Trek. But the speed of light limits our travel on the physical LOR. If aliens are present on the Earth, they surely have a better method of travel than the use of chemical rockets.

In our everyday lives, science and technology have brought us such things as microwave ovens, Velcro, digital cameras, remote controls, laptop computers, cell phones, and so many other things that it would be almost impossible to list them all. We like having such things in our lives. They give us more control over our environment and give us a sense of stability, but the degree of control we have over our physical and psychological health is even more important to our wellbeing.

Modern doctors have long lists of diseases, tests, symptoms and pharmaceuticals. Even so, if you have a disease, you often find that the doctors guess at your illness, even after multiple expensive tests. More often than not, they treat the symptoms rather than the disease. Antibiotics are prescribed even when there is no known bacterial infection. The deadly strains of cancer are little-understood, often diagnosed too late, and the treatments are destructive not only to the cancer, but to the remaining health of the patient.

I don't mean to paint a black picture of modern medicine. Medical researchers have accomplished much, virtually wiping out some diseases such as polio and tuberculosis. The treatments for many other diseases have improved dramatically. But even with all these advances, I contend that medical science is still an infant. Perhaps some years from now when genetic research has led us to a deeper understanding of the human organism, our doctors will be more successful in diagnosing and treating our maladies. However, it would be more beneficial if our medical science and technology could help us avoid ailments rather than treat their symptoms after we become ill.

Psychologists have even more difficulty recognizing and treating psychological maladies. While a medical doctor can use a variety of invasive techniques to actually look inside the body to diagnose, and sometimes to fix, physical problems, a psychologist doesn't know where to find a person's subconscious. They have to hope that the observable reactions of the patient are indicative of the malady. If the psychologist is able to accurately diagnose the problem, she or he seldom has a standard, well-understood treatment to solve the problem. The field of psychology has multiple theoretical systems categorizing and proposing solutions for psychological problems. While medical researchers often have the option of observing the reaction of a disease (e.g. bacteria or a virus) to a particular drug under a microscope, researchers in the field of psychology cannot directly observe the reaction of a subconscious to a particular treatment. Again, they have to rely on behavioral observations.

To summarize, I think we can agree that the fields of mainstream science and the associated technologies have developed rapidly, and perhaps the rate of development is increasing. The effects on life on Earth, both for humans and for other species, has been tremendous, some for good and some effects are not so good. But until now, the development has been limited to the physical LOR. What might life be like if mainstream science were not restricted to the physical LOR? I wonder ...

Chapter 10: A Peek at Tomorrow

On this leg of our journey, let's see how our lives might be affected if our science and technology included nonphysical LORs. What if a version of the present book had been written fifty years ago, and mainstream science had eagerly embraced the previously unexplored areas of nature at that time? How might our lives be different today? To answer these questions, I'm going to tell you a fictional story of a day in the life of a family in that hypothetical present-day. This is of course completely conjecture, but I hope it will provide some sense of the total impact on the life of a family, rather than just looking at the contributing technologies, one by one. I realize that fifty years of research would develop much more than I will include in this story, so please take it as only a partial indication. On the other hand, I will include a subset of ideas and principles that are commonly accepted by those who have been practicing etheric awareness and astral travel for hundreds of years. Remember that many people in Europe and throughout the world began to show interest in nonphysical LORs—again—immediately after the Inquisition became non-fatal in the middle of the 19th century. I'm not claiming that their conclusions are valid, but I am including some common forms of their conclusions in this story for your reference.

Bel awoke long before the alarm went off, but she didn't stir. She lay quietly, trying to think about her patients, trying not to think about Paul's send the previous night (telepathy is common). She knew he had been angry, and he probably didn't mean it, but it had cut her deeply, especially the image of the squealing prostitute. She thought (but was careful not to send), *Mr. Sherman could sue the health center. I have to document his case very carefully.* Her mind came back to Paul. She couldn't help feeling guilty. *I should have just refused.* She was careful to keep a blank wall toward her husband, not sending her emotions or thoughts. She forced her thoughts back to her work. *And then there's poor Mrs. Whittier. She's not smart enough to be able to manipulate her own transit energy, but that's no excuse. Doctor Fowler should be helping her. I have to do something about him.* Her more objective side argued that she had tried to dissuade Paul, but he had insisted. Her anger surged, but she kept it contained. *I can refer Mrs. Whittier to Doctor Nelson. I'll call him this morning and explain the situation. I'm sure he'll take care of her.* Her anger had died down a little. *I have to find out how serious he was. I thought we were doing OK.* Pang of guilt. *I thought he was getting what he needed. I wonder how long he's been dissatisfied.* Fear intermingled with the guilt. *I don't want to live without him.* A tear trickled down her cheek. In control again. *I'll speak with the director about Fowler. I will not allow him to mislead my patients!* Disproportionate anger toward Doctor Fowler, but still completely contained.

As a physician, Johns Hopkins had trained her to send only those thoughts, emotions and feelings that she deemed appropriate. Her professors had told them stories of malpractice suits based solely on inappropriate telepathic communication with patients. She was an M.D., but not a specialist, like most M.D.s today. She had chosen the path of a general practitioner to treat a wider range of patients. Her deep caring for people was like a neon sign—or a halo. Patients flocked to her, even though she only worked part-time. They were willing to wait weeks for an appointment to see her, even though they could walk into the Center and see another G.P. without an appointment. *I'll brush up on handling transit energy. Maybe I should try helping Mrs. Whittier before I call Nelson. No. He has specialist info and experience that I couldn't match. I'll call him before I go in this morning. That would be best for Mrs. Whittier.*

The alarm chimed and she felt her husband bolt out of bed. *I gather he wasn't sleeping, either.* Anger. *I hope he's hurting!* Fear. *I hope he doesn't leave.* She caught herself just short of tears and got out of bed. Putting on her bathrobe, she stepped down the hall and knocked on Anita's door. "Good morning, sweetheart. Up and at 'em."

There was a muffled "I'm up, mom. Good morning."

Anita was twelve. Bel thought, *She's such a sweet girl.*

Another couple of steps and she knocked on Nate's door. "Good morning, honey. Are you up?"

He said, "OK!"

She thought, *having a sixteen year old boy is enough conflict for most families, but Nate's squares and other aspects are indicative of his problems. I'm glad Bess is counseling him—and us. All psychologists know a great deal of astrology theory and practice. It would make no sense to be a psychologist without such knowledge. Luckily, Bess specialized in personality problems associated with aspects like Nate's. We have an appointment with Bess next Tuesday. I hope he'll be OK until then. Next year, we should be able to get him to work through some of his past lives and overcome most of his problems.*

She thought about Bess and their history together. *Doctor Bess Morehead. She's trained to enter a client's subconscious and investigate the root cause of the problems indicated in a client's astrological chart, including past lives when applicable. It wasn't ethical, and I'll always be grateful, the way she helped me visit Nate's most relevant past lives. Without that knowledge, we wouldn't have made it this far as a family. I wish I could have shared it with Paul, but that wouldn't have been fair to Bess. Bess and I have been best friends almost since we met at Johns Hopkins University. She even chose to practice here in Denver to be near me. I could never betray her. Never.*

As she turned to descend the stairs, Bel remembered the most difficult of Nate's past lives and cringed. *If I could share that with Paul, he would be more understanding. But if anyone ever found out that Bess showed it to me, she would lose her medical license. The laws governing privacy are very strict. It wasn't that long ago that people invaded each others' privacy, unregulated. People can be so cruel. The laws are necessary. Maybe in the future, our laws will have room for special cases, like ours.* She sighed. *Being a buffer between Paul and Nate is no fun. Maybe next year Nate will get it worked out. God, I hope so. . . . and I hope Paul will still be here.*

I wonder how much Nate had to do with Paul's reaction last night. Sigh. *No, it's me. Just me. I've got to talk to him about it. I'll find a way to get him alone. I have to.*

She walked slowly down the stairs, feeling old and tired and sad. Collecting bread, eggs, milk from the refrigerator, she began to fix breakfast.

Nate came into the kitchen. *Good morning mom.* With that, he sent her love and an image of spring flowers resplendent not only in colors, but in fragrance. *I wasn't quite awake when you called.*

She sent, *Bad dreams?* Image of a giant version of Paul with a huge fist in Nate's face. That had been Nate's recurring nightmare. In one of his more difficult past lives, his father had beaten him frequently until one day, the beating went just a little too far and that earlier Nate had died in agony from internal injuries. Of course, she couldn't tell Nate that she knew, and Bess assured her that it would do Nate no good to hear about his past lives—he had to remember them and re-experience the traumas. Only then could Bess lead him to rework his reactions, thus changing the patterns in his personality.

Nate said, "No. I have an Algebra midterm today. I studied until midnight, and then dreamt about it. It was awful. I had numbers and formulas all around and I was drowning, over and over." Then he sent an image of the dream, complete with feelings.

"Oh, I'm so sorry," she said. "That *was* awful, but I'm sure you'll do fine." She hugged him and kissed him on the cheek.

He smiled smiled and sent, *I'll finish the eggs and toast. You go get ready.*

She and nodded. *Thanks.*

Paul was coming down the stairs as she went up. He said, "Gotta run. I've got some serious shit to deal with at work."

She stopped and watched him hurry out the door without breakfast or even a kiss. A tear came to her eye as she thought of their normal routine, frequent random kisses and breakfast with the kids, everybody talking about their plans for the day. She slowly trudged up the stairs.

Showered and clothed, Bel joined her children at breakfast.

Anita sent, *Where's dad?*

Bel said, "He said he had something important he had to get done at work." She didn't dare send anything. She knew any message would be tainted by her emotions.

Anita seemed to understand and fell quiet.

Nate sent, *Last night, he told me he's got a mess. He wouldn't tell me the details, but he said he might get in trouble.* Image of his father punching another man in the mouth.

Bel said, "Oh, I'm sure your father wouldn't do that. He didn't tell you anything else about it?"

"No. I had to go study, remember?" Image of himself surrounded by numbers and formulas, drowning him.

Bel laughed. "I remember."

Anita yelled, "I see the bus!"

Both children jumped and, grabbing their backpacks, ran out the door.

Bel sat alone at the table and thought, *I wonder why he didn't tell me? Should I be more worried, or less worried? If he's not talking to me, we may have a bigger problem than I realized. It's not useful for me to think about it now. I really need to talk to Paul!*

Quickly gathering her things, she headed for work. She called Doctor Nelson's private number from her cell phone.

He answered cheerfully, "Hello, Fulton. What can I do for you on this beautiful Colorado morning?"

"Hello and good morning. I would like to refer a patient to you. Could we talk now? Is this a good time for you?"

"No worries," he replied. "Something special?"

"Yes, I'm afraid so. It's a bit sensitive. You see, I referred a patient to Doctor Fowler, and . . ."

"Let me guess. He insults your patient's intelligence, is condescending and doesn't help. Am I close?"

She hesitated, surprised. "Umm. I guess I'm not the first doctor to see this."

He chuckled. "I guess not. The only reason he hasn't had malpractice suits is that he knows how to hoodwink the patients, and the doctors who take over from him haven't wanted to press it. He's not exactly a credit to our profession."

She sent anger and said, “That stops now. I’ll deal with it.” Sending a feeling of gratitude, she said, “I hope you don’t mind helping my patient.”

He sent mirth and said, “Of course not! What do I need to know?”

“My patient’s name is Mrs. Whittier. She’s not too bright, and she can’t understand how to redirect and use the energy from a major Mars transit. The influence is at a level seven.”

“When can I see her?” A sense of urgency.

“She has an appointment with me at 10:00.”

He said, “I’ll make room for her in my schedule.” Image of a woman hanging by one hand from the edge of a cliff, with his hand reaching to pull her up.

She hung up the phone and sent an image of a big smile, with the lips mouthing, “Thanks.”

As Anita ran toward the curb, the bus stopped in front of their house and dropped softly to the street. She found a seat with her friends and they began to chatter. Reaching the school, she saw Nate get off the bus before her and she hurried to catch up. “Hey Nate! I’ve got a question.”

“You’ve always got a question.”

Undaunted, she asked, “What makes the bus go?”

“That’s pretty advanced for you. I’ve seen some stuff about it in books, but I won’t cover it in class until next year when I take Physics.”

“OK, but you know something.”

He grinned. “All right. Cars and trucks and buses all used to ride on wheels that were turned by an engine that burned oil. You know what oil is.”

She nodded.

“About 40 years ago, they figured out how to make machines that are etheric and physical. Like us.” He pointed to her aura and to her physical body.

“That seems obvious. What took them so long?”

“We’re going to be late for class. Ask you history teacher about it. And ask your science teacher about the bus. If they don’t tell you enough, we can talk about it after school. OK?”

“OK.” She sent love and gratitude to Nate.

He smiled and ran.

She ran to her locker and then to her first class—history.

Anita took her usual seat, four rows back and two seats to the right of the center aisle. As the class began to settle, she raised her hand.

Mrs. Dauer said, “Yes, Anita?”

“What took them so long to make machines that are etheric and physical?”

"I think you mean machines that have an etheric part and a physical part, connected together. The answer is kind of long, but I was planning to teach part of it today, anyway. Remember what the last era was? Christians—and I'm a Christian—believe that Jesus Christ was born a little over two thousand years ago. Throughout most of history, people thought that the gods controlled everything. They didn't understand nature, and they lived in fear."

Anita interrupted. "What were they afraid of?"

"They didn't understand very much, so they thought the gods were fickle—fickle means they were afraid the gods would change things whenever they felt like it. In other words, they were afraid of everything. And in their fear, they concentrated only on physical things."

Bobby asked, "But Mrs. Dauer, how could they do that? Everybody sees etheric stuff—well, almost everybody." He glanced at Fred, who quickly looked elsewhere.

Mrs. Dauer said, "Now, don't you start. Some people can't see the etheric level of reality. It makes their lives more difficult, like people whose ears or eyes don't work correctly. They deserve just as much respect as you do."

Bobby said, "Yes, ma'am."

"Now, where was I? Oh, yeah. In those days, not many people admitted to being able to see the etheric or astral levels, and everyone made believe that they didn't exist, except the church. It's too complicated to explain now, but when modern science got started, hundreds of years ago, the scientists were too frightened to admit that other levels of reality existed. They continued to deny it until the beginning of the modern era, fifty years ago. And what do we call the modern era?"

The students spoke in a jumble, "The Open Times."

"That's right. The Open Times began when scientists accepted the existence of other levels of reality and began to study them. Our physical science was already well-established, so they were able to add lots of nonphysical knowledge quickly. So, Anita, to answer your question, it didn't take them very long at all to begin creating composite machines. Does everyone know the word, composite? Good."

Anita said, "My brother said they didn't figure it out until forty years ago. But they could see that we're composite—things. Why did it take them so long?"

Mrs. Dauer smiled, "Hmm, yes, I guess it did take those scientists a long time to do their research and then develop technology based on what they learned. Yes, ten whole years. I see your point."

Anita sat back in her seat with a smug smile, thinking, *I was right. I'll have to explain it to Nate.*

Nate asked, "I'm not clear on something. Why aren't we allowed to enter most of the inhabited star systems?"

Mr. Collins replied patiently, "The advanced civilizations have made it very clear that humans will not be allowed to interact with their worlds except under well-defined conditions, as and when they specify the conditions."

"Yes, I know, but why? Are they afraid of us? Are they concerned that we will learn too much? Do they think we'll steal things? Do we even know why?"

Mr. Collins took a deep breath and sighed. "I understand your curiosity, and I wish I could answer your question, but honestly, we can't be sure what their agenda is. All we know is that we are allowed to travel widely in the galaxy with the exception of marked star systems. If humans enter those systems, they are expelled or killed."

"How are those systems marked?"

"I understand that it's a barrier that our ships sort of bounce off of. You know people. Some have been persistent enough to force their way through into the forbidden systems. Usually, the aliens killed them and made a point to let us know by returning the ship with everyone dead."

Frank Johnson asked, "Why haven't we done anything about that?"

Mr. Collins smiled faintly. "Like what? Their science is millions of years more advanced than ours. We're very lucky that they allow us to leave this planet at all. Besides, if you entered a military installation here on Earth and you were shot and killed, your family would have no recourse. If you go some place you shouldn't be, you suffer the consequences. That's all there is to it. Now, let's get back on track. This is a social studies class, and this week we're covering the impact of interstellar travel on society."

Nate interrupted again, "How many alien races are there, and are we going to study some of them?"

"I've read that scientists suspect there are over a million inhabited planets in this galaxy, with roughly thousands of distinctly different races. For the time being, we can't investigate, and we know very little about them. The only ones we really know anything about are the six races that have approached us. And we only know what they want us to know. They have assured us that when we are more mature, we will be included in the interstellar community."

"When will that be?"

Mr. Collins looked at Nate and shook his head. "Is that a serious question?"

"I guess not. I was just hoping somebody knows the answer."

Shaking his head, Mr. Collins said, "To continue, space travel became practical about 25 years ago. In the next five years, we explored enough of our galaxy to find some planets that we would be allowed to colonize. As you know, our ships can travel to most of the systems in our galaxy in less than a week. The impact on our civilization has been dramatic in deed. We have established colonies on at least a hundred planets so far. Our ships primarily carry people, machinery and information. Few colonies have allowed any plant or animal transport. They depend on the local ecosystems, and our experience in places like Australia has taught us not to introduce plants or animals that are not native. . . ."

Nate interrupted, "What do you mean about Australia? Do you mean here on the Earth?"

Nate noticed the annoyed expression on Mr. Collins' face as he cleared his throat. "Yes, Australia here on Earth. Check it out on the Internet. Look at what happened in Australia in the early to mid 1930's, with the introduction of new species. Now, Mr. Fulton, I would like to continue."

Nate smiled at the rebuke and leaned back in his chair.

"As I was saying, our fleet of interstellar ships is growing rapidly. Some of those ships are used strictly by scientists to explore and to investigate astronomical phenomena. But our interest in this class is on the ships that service our colonies. Our ships make runs to and from most of our colonies at least once a week. The waiting lists for passengers are months long, and growing. Most of the colonies are reported to be exquisitely beautiful, and life is usually described as idyllic.

"Now, many of our colonies are being established for specific groups. Jews are populating New Israel. There is a New Africa, Spanish colonies, white colonies, Catholic colonies, artist colonies, and so on. Most of the colonies encourage tourists, but only the "right" people can become citizens. Remember the social development on the Earth before The Open Times. Prejudice was a major theme when dissimilar peoples desired the same resources, such as land, wealth, minerals, etc. In more recent years, we have found that interracial marriages have resulted in a more integrated society, with less conflict. However, with so much explicit segregation in the colonies, some sociologists predict the result will be wars. But others believe that the other interstellar races will not allow interstellar conflicts on that scale, and that the human race will simply become splintered, with a reduction in wars everywhere."

Nate asked, “What effect do you expect it to have here on the Earth?”

“Conflict is dwindling on the Earth because groups that used to fight are moving farther apart. The population on the Earth is 10% less than what it was at the beginning of the current era. People have less to fight about. With the dwindling population they have more room, and we are getting natural resources from other worlds, often in this solar system, like the mines on Mars. The rate of domestic violence has also decreased dramatically since we began to move our prison inmates to New Alcatraz. Most of the criminals disperse on that planet and refuse to come back to the Earth. I suppose it is remindful of the British export of their criminals to Australia, but New Alcatraz is reported to be a beautiful planet and life there is supposedly easy. Some of our more conservative-minded people think it is far too good for criminals, but the positive impact on our society is undeniable.

Mr. Collins glanced at the clock by the door, and began to speak rapidly. “Tomorrow we will cover Chapter five, the evolving social structure . . .” The bell rang. “. . . on New America. Read it. Dismissed.”

Nate jumped out of his seat, exiting the room with Frank Johnson. Frank leaned over and whispered, “I think we were getting to him.”

Nate just laughed. A few steps down the hall, and they were in Mrs. Apple’s room. She had written on the board, “Science Class” in big letters. While Frank distracted her, Nate changed the last “s” to an “h”, and then took his seat. Smiling, Frank joined him in the next seat. Nate waited expectantly for Mrs. Apple to notice the Science Clash.

Paul turned his car toward work. He would be early getting there, but he couldn't be with Bel and act as if nothing had happened. *I really messed up.* He hadn't slept at all. *I don't think she'll ever forgive me. I was so stupid. She said she wasn't feeling good—tired and upset about her patients and other doctors, or something like that. Then I insisted on sex. Sure, my job's up in the air, but that was really stupid.* He began to think again about George at the office. George was trying to make him look bad. He was sure that George had done the same thing to others, but he couldn't prove it. George was trying to take credit for one of Paul's ideas. The mental privacy code did not allow an employer to check directly, so their manager had to rely on physical-level evidence. Paul had no evidence that it was his idea, while George had immediately documented Paul's idea when he first described it in a private meeting. *I won't let him get away with it. I don't care what it takes.* The idea was a new type of data interface between a physical computer and an etheric-level communications device. He was sure the new interface would simplify their programs and it would generalize to other applications. *I* am *going to get credit for it!*

Stewing about it as he walked into the office, he approached his manager's office. Ralph looked up and said, "Fulton, just the person I wanted to see. Come in." Managers were required to use verbal communication to avoid telepathic eavesdropping. Ralph closed his office door and sat down across from Paul. He described the interface idea to Paul as it had come to him in an email from George. Then he said, "Paul, was that really your idea?"

Paul nodded. He could feel his face twitching, and he didn't think he could trust his voice, so he didn't say anything.

Ralph said, "Please calm down. We'll work this out. Let me make a suggestion. If you don't like it, we'll figure something else out, but this is what would work best for me."

Paul stiffened just a little. *What he's telling me is that I'd better like his approach or I'll pay for it later.*

Ralph continued, "I'd like to give both of you credit for the basic idea, as a team. As soon as you get back to your office, I want you to start working out the details. Can you give me a more detailed version of the interface later today?"

Paul felt his stomach cramping and his back was so stiff he could barely move. His face felt like a stone mask. He nodded slightly.

Ralph hurried on. "Good! I don't want you to meet with George again until you complete the interface description in detail. I want updates on the interface every day until you have all the principles and ideas and details worked out. Then I want you to put it all in a presentation. I'll arrange for you to give the presentation at an executive staff meeting. If we work this right, your name will be permanently branded on your interface. As to George, let me deal with him. OK?"

Paul slowly relaxed as he realized that Ralph was going to help him. After a full minute, he said, "I can't tell you how much this means to me. I owe you a big one. Thanks!"

With a grin, Ralph said, "Now get out of here and get some work done."

Still slightly dazed, Paul moved toward his desk. The more it sank in, the better he felt. Ralph was taking care of him. He was going to be OK. By the time he sat down, his face held a smile. He thought, *I may not be able to wipe this grin off my face for a week.* But then he remembered his screw-up with Bel and he almost cried. He called her, but she was on her phone. He left a message asking her to call him.

He tried to reach her several times that morning. He left messages, but she didn't call back. He had hoped to meet her for lunch, but he knew how busy she was.

Just before lunch, he was returning from the rest room when Ralph waved him into his office. "Let's go out for lunch. I'm buying."

Paul reluctantly agreed. As he was leaving Ralph's office, he saw Tony Rupp enter the office. Paul had known Tony since high school. They had been in track together. Tony had been nearly Olympic in the half-mile event, beating Paul consistently. At first, Paul had been incensed, and did everything he could to win against Tony. He had changed his diet, he used weight training and he ran every day. But eventually he had to admit that Tony would always be faster, and by then they were best friends. Even last weekend, they went hiking together near Deckers, south of Denver.

As Tony approached, tears came to Paul's eyes. Tony liked to joke and play. Tony had entered through the front wall, but this wasn't the sort of joke that he would play. Paul sent, *What happened?*

Tony sent, *Accident on I-70.* Image of several cars piled up, with Tony in a mangled mass of steel and plastic.

Paul sent, *Maggie?*

I went to her first. She'll be OK, but she's pretty upset.

I'm not very happy about it, either. Why the fuck couldn't you wait a few more years!?

You think I wanted to die today!? My old friend, I don't want to leave, but . . . well, it just happened. I'm sorry. We had some really good times together, didn't we?

I'm the one who should be sorry. Here I am giving you a hard time, and you're the one who died. From what I've heard, we'll probably get back together. Will you be all right? Is there anything I can do?

I'm sure we will be together again. Actually, I asked one of my guides about that and she said I will be coming back, and as much as you and I love each other, we couldn't be kept apart.

Paul sent, *Will it be OK if I think about you often? And all the good times we've had together?* Image of a very surprised high school teacher when they had left a garter snake in his desk, painted like a rattlesnake.

Tony grinned. *Please do. I know I'll be thinking of you.*

Ralph sent, *Tony, I'll take care of Paul while you're gone. He's a good guy. So are you. Judging by the beings around you, I'd say you'll be well taken care of where you're going. I'll have a job for you when you get back and get old enough.*

Tony laughed. *You take care, too, Ralph. It's been nice knowing you. Goodbye for now.* Tony left through the ceiling.

Ralph suggested, "Paul, why don't you take the day off? Go hike in the mountains or something. I know it's important to you, but you can get started on the interface development tomorrow. You've just lost your best friend. You need some time to yourself."

Tears in his eyes. "I know, but not yet. I need to concentrate on work. Maybe I'll take tomorrow off."

"Something else is bothering you. Would it help to talk about it? Come back in and sit down."

"No. I can't. Let me get to work."

Paul turned and walked back to his desk, feeling like he'd just been kicked in the balls. Arriving at his desk, he ran for the restroom and threw up. His stomach still cramping, he thought, *Hiking? Alone? Today? I could join Tony. No! Bel. Anita. Nate. Bel may . . . can't think that. Work.*

Back at his desk, Paul turned his computer on and stared at the screen, trying to force his mind to work. After a few minutes, his anger toward George helped him to concentrate on the new interface, and he began to draw diagrams.

Bel said, "Mr. Sherman, the tests verify that your colon cancer is at stage II. It is becoming dangerous, and we should operate as soon as possible."

He just nodded his head.

“Could you tell me the dates and locations of your checkups in the last twelve months?”

Mr. Sherman smirked and chuckled. “You’re concerned I’m going to sue you. I’m not. I just didn’t feel like coming in. The truth is, all this psychic stuff gives me the willies. When I grew up, this stuff was considered to be pseudoscience at best, and usually fraud. Yeah, I know what the books say, but I don’t quite buy it. I don’t think everybody was wrong.”

“Have you read any of those books?”

“No. I didn’t see any point to it. I already know it’s wrong, so why read it?”

“But Mr. Sherman, your cancer was found using the very technology that you think is wrong. How do you justify your point of view?”

He said curtly, “I don’t need to justify my beliefs to you. Just schedule the damn operation.”

“All right. I have a couple of papers for you to sign. This letter states that we were not responsible for the late diagnosis of your cancer. And this one gives us authorization to perform the operation.”

As he signed them, he said, “I suppose if I didn’t sign both of them, you wouldn’t give me the operation.”

“No, we can only require you to authorize the operation.” But he had already signed both forms.

Bel collected the forms and said, “Thank you. Just stop by the reception desk. They will schedule you for the operation. And thank you for coming in today.”

“Yeah, thanks, Doc. And I don’t blame you for participating in that psychic shit. You don’t have any choice, ‘cause it’s what doctors do now. It just bugs me.”

She nodded and he left her office.

Mr. Sherman had no sooner left than Doctor Nelson stuck his head in. “So how did it go?”

“I dread treating people from that generation. He still believes that we were all misled and that we should only be using physical technology.”

“Even though the new technology is saving his life. I’ve got my share of those patients. How old is he?”

“Seventy one.”

He whistled softly. “Yup. People of his generation were taught to believe that anything nonphysical belongs to the church. Even so, most of them have adjusted. Most have even taken the classes to develop their own awareness of other levels, although not many of them have been able to develop telepathy, like the rest of us. Sometimes I feel sorry for them. They must feel so left out.”

“I agree. Speaking of which, how’s it going with Mrs. Whittier?”

"Like you said, she can't manipulate the influences herself, but with her permission, I adjusted them to reduce the risk. I've set up weekly appointments for her until the transit is over. She'll be fine."

"That's good to hear. Thanks! I owe you one."

"No you don't, but you're welcome anyway. Please let me know if there's anything else I can do."

She said, "Thanks!"

Bel's next patient arrived, and Doctor Nelson fled to the hallway.

She thought, *This one's easy. I see from her aura, it's just a cold.*

A few minutes later, she was talking to another patient. "Mr. Jackson, I've read your records and looked at your natal chart. It looks like you have a predisposition to heart trouble. From what I can see, you don't have an infection, so I'm going to refer you to a heart specialist."

"I'm just feeling a little under the weather."

"Has anyone mentioned your predisposition to heart trouble before?"

"Yes, of course, but maybe I have a cold or the flu or something."

"Mr. Jackson, if you had a cold or the flu, I would be able to see that in your aura. And other types of infections usually have other symptoms, such as a rash. I strongly recommend that you see this specialist."

"What will he do?"

"That will depend on what he finds, but I know he normally begins with a non-invasive internal scan. What he sees will determine the next step. He will explain it all to you. Are you OK with that?"

"Yes, of course. Thanks for your help, Doctor Fuller."

"You're welcome. Please stop by the reception desk. They will schedule an appointment for you."

As she was writing up her notes for the heart specialist, Ms. Patterson entered her office. Bel said, "Please have a seat. I'll be right with you." She quickly finished writing and retrieved Ms. Patterson's records.

"Your notes are here from the RV specialist. He found a small lump in your left breast, which he believes is stage-I breast cancer. It's nothing to worry about, if we operate within the next month. It can be removed completely without leaving a noticeable scar. I am referring you to an excellent cancer specialist. She will handle your case from here on. Besides dealing with your current cancer, she may be able to tell you how to reduce the risk of getting cancer again. Do you have any questions for me?"

"Thanks, Doc."

As Ms. Patterson left, Bel thought, *My next patient won't be quite that easy.* She noticed the message light flashing. It was a voice message from Paul. Her face grew warm as her anger surged. *Let him sweat!*

Ralph said, "I could see how important your interface idea is to you, and I agree. If you do a good job on the details, you may get your own group. But I still think you need to take some time."

Paul replied without looking up from the menu, "I know you've got my back, and I appreciate that. And, yes, there is something else, but I'm not ready to talk about it. Maybe I never will."

"I understand, and I won't pry, but I'm here if you'd like to talk. In the meantime, we can focus on work."

"Thanks. I appreciate that."

"As I was saying, I can see how important your idea is to you, and I don't disagree. But I don't want you doing anything that might mess up your career. That's why I'm going to tell you what I plan to do with George. If you ever tell anyone, I could get in big trouble. All right?"

Paul nodded, all his attention on Ralph.

"As you know, I can't ask a company psychologist to dig into the matter. It wouldn't take much for one of them to get at the truth, but the privacy laws are very specific."

"Yes, I know. The law says you can only pry into another person's thoughts if a serious crime has been committed, and then only with a court order."

"Precisely. But there's nothing says I can't set a trap. I'm going to pair George with a mild-mannered engineer by the name of Janise Peak. Do you know her?"

"I've heard she's fairly sharp."

"Even better—she's very creative and she covers her ass better than anybody. . . . but she doesn't advertise that little fact, and I hope you won't mention it to anyone."

"You kidding? I'll go to the grave with that secret."

"Enough said, then? Are you satisfied?"

"Just one question. What can you do to George if you catch him stealing her ideas?"

"*When* I catch him, someone will leak it to the biggest gossip in the company. Then I'll fire him. See my point?"

"You're an artist."

Ralph sent a caricatured image of himself, patting himself on the back while simultaneously rubbing his knuckles on his chest.

Paul laughed so hard that he cried. When he had settled down a little, Paul said, "Thanks. I needed to hear that."

Ralph nodded.

The waiter arrived and they ordered lunch.

After the waiter left, Paul said, "I haven't taken many of the management classes, but as I understand it, you use astrology and aura-matching as inputs when you create a team. Why did you match me with George?"

A sly smile came to Ralph's face.

Paul said, "You bastard! You already knew about George."

"I suspected, and I was hoping that you would give me proof. But, noooo. You just laid it out for him and made it easy for him to steal your idea. Even so, I'm protecting you. You'll get the credit. Don't worry."

"Why didn't you give me a hint or something? I could have covered my ass if I'd known George was going to bite it."

"Sorry. I couldn't tell you my suspicion. That would have been inappropriate. All I could do was watch and make sure you didn't get hurt in the process."

"I suppose so. All that privacy bullshit."

Ralph shook his head. "It's not bullshit. We were required to read about the misuses of nonphysical knowledge in the early part of the Open Times. Some of the things people did with astrology data, aura characteristics, past-life knowledge, etc. The misuses were as ugly as the constructive uses are beautiful. The laws are well-justified. I think we should be happy that we're allowed to use nonphysical knowledge as much as we do. Seriously. Our new knowledge has helped improve the work environment in more ways than you might suspect."

"I know. I've done some reading, too. In the old days, a manager had no real basis for selecting team members, and the level of conflict in the office was a lot higher than it is now. Now, we tend to like our coworkers and we're more likely to look forward to coming to work each day. But you . . ."

“Yeah, I know. I won’t do anything like this to you again. I’ll make sure you have compatible teammates. And I’ll get you through this one, looking better than you did before.”

“Promise?”

As always, Bel took a few patient files with her when she left the health center. They were patients she would see on the following day. As always, she arrived a few minutes before Nate and Anita got off the bus.

The children were arguing as they opened the front door.

Anita said, “Mom! I can’t get Nate to admit that he was wrong.”

Bel hugged and kissed each of them before she responded. “All right. What are you arguing about?”

“Ten years is a very short time,” Nate said. “It took our scientists ten years to figure out how to make composite machines, like the buses, that use etheric technology. I asked my science teacher and he verified it.”

Anita said, “I’m just saying that ten years is a long time to do something that obvious. After all, we’re—composite—machines, and they already knew that.”

“Honey, sometimes, even when you know part of the answer, it takes a long time to solve a problem. And time is relative. I know ten years is a really long time from your point of view, but those scientists were very proud when it only took them ten years to build the first modern automobile.”

"Humph! All right, but my history teacher agreed with me."

"You're both right."

Bel, watching Nate's smile, added, "Nate. Think back to when you were twelve, and keep in mind that you still have a great deal to learn about the world."

He said, "Fine. You were right, too."

The children found things to do, leaving Bel to read the patient files. As she was opening the third file, Paul drove into the driveway and his car settled to the pavement. She met him at the door, concerned. "Why are you home so early? Did something happen at work? Are you all right?"

"Tony died in an accident on I-70 this morning."

She thought he would say more, but Anita ran to him and gave him a big hug. "Hi daddy! Why are you home so early?"

"Tony's dead. He came to the office and told me himself. I miss him already. Ralph wanted me to take the day off, but I had to get some work done first."

Nate came into the room and hugged his father. "Dad, I'm sorry to hear that. Are you OK?"

"Yes, I'm fine. It was clear that Tony will be fine, and we agreed that I'll be seeing him again."

Anita asked, "Are you talking about re-in-car-nation?"

"Yes, honey. He said he'll be coming back. But we don't know when or where or who he'll be. But we think we'll meet again no matter what."

Nate asked, "Is it possible he'll incarnate on another planet, like maybe one of our colonies?"

Bel said, "Sweethearts, please let your father be. He has enough on his mind. Give him a chance to relax. Maybe you can talk more, later. OK?"

Anita said, "Daddy, I love you. I'll go now."

Nate said, "Sorry, Dad. I'll catch you later."

Both children hesitated by the door.

Bel said, "I'll give you a chance to relax, too. OK?"

She could see in his eyes that he understood.

He said, "Before I relax, I want to deal with Pan. I saw him in the next block in a tree. I think he's stuck up there, again."

Nate said, "Let me get him. You relax."

Paul said, "No, I can use the walk, and it will help me relax. You kids go on and play. Let me spend a few minutes alone with your mother." Turning to Bel, he asked, "Will you come with me? Keep me company?"

She nodded. "I'll get the carrying sack." As she retrieved the cloth sack from under the sink, she remembered making it. Paul had fallen out of a tree, attempting to carry Pan to the ground. Luckily, he had only sprained his ankle, but she vowed that he would always have both hands after that. She inspected the carrying strap and put it around Paul's neck.

As they walked to the street, she said, "I'm truly sorry about Tony."

He responded, "I'm more concerned about us."

"Mhmm." Hearing her own voice, she was a little surprised at how cold it sounded. Her face became hot as various emotions raced through her—anger, fear, guilt, compassion, love, anger and guilt.

They walked in stiff silence to the base of the tree.

Pan let out a yowl as if to say, "I'm here! Here I am!"

Bel, pointing to the right: "I think there's a way up this side of the tree."

Paul began to ascend as Pan meowed plaintively.

Her back stiff, holding back tears, she watched as he approached the cat. She had designed the cloth sack with a large opening, and they had trained Pan to enter it. After all, she thought, *He's had plenty of practice. We've had to rescue him about once a week.*

When Paul had secured Pan, Bel began, "Paul, . . ."

Without moving from the limb, Paul said, "I know. I know. It was the stupidest thing I've ever done, and I'm really really sorry. I don't know if you can ever forgive me. What can I do to make it up to you?"

"Why . . ."

"Honey, yesterday Ralph forwarded an email to me. George documented my idea for the new interface I told you about, as if he had invented it. I knew it would be my word against George's, and I was so upset I couldn't talk about it."

"Get your ass down here."

Paul, began to descend slowly: "I made up my mind that I would either get the credit for the interface or I would do something to George and then quit."

A few feet lower, he was quiet for a second while making a dangerous move and adjusting his grip: "I know I should have listened to you last night. I should have let you alone. But I thought maybe having sex would relax me some. I'm really sorry."

Hesitating again, he made another dangerous move. "Then when you didn't respond, you just laid there, I blew up. I sent you that awful picture of me with a prostitute and her screaming with joy. I'm so very very sorry. I was so mad at George. . . ."

Bel knew enough psychology to understand his reaction. *All right, so now I know where the image came from. It was partly Paul's anger at George and his fear for what he would do and for what would happen to him. But that's not the whole story. Sweat, you bastard.*

Paul, didn't hesitate in his monolog. ". . . Then I was so scared. I almost forced you to have sex and then I did that send. I felt like I'd blown everything. I was going to . . ."

Bel finally took pity on him and sent love and calm, surrounding him with it.

". . . lose the love of my life and my job and my career all at the same time."

He stopped on a limb near the ground and visibly relaxed. "Thank you. . . . But my send was unforgivable. With the picture, I sent that you should have a TV on the ceiling and that a prostitute would act like she cared. I didn't mean it, honey. It was George."

"Paul, we'll work it out. Come on down."

Reaching the ground, he held her by the shoulders. She didn't move, still angry. He said, "I don't know how you could ever trust me again. Can you?"

"I thought it was my fault. I didn't get much sleep. . . "

Paul, interrupted, shaking his head, "No. No. It was me. I was so messed up I didn't sleep at all."

"No, I thought about it all night and all day. We need to work on our communication. We've let it slide. We have to talk more about what's important to each of us. And we've got to be better in bed. Don't deny it. The more I thought about it, the more I realized I haven't been very exciting for you. Umm, and I need some things, too."

"Yeah, I know. When we made love, I used to send you all those sexy images and share my emotions with you. But that was before Nate was old enough to know what we were doing."

"And I used to send and share my ecstasy. Our lovemaking was incredibly intense and wonderful. Honey, . . ."

"I know, it's my fault. You said the children would be all right if we kept sharing like that, but I stopped. I'm so very . . ."

Bel put her finger to his lips. "Honey, I stopped, too. I let our lovemaking become secondary to our careers and to our children. We let it slide too far. We need to talk more, and we need to send when we have sex. I forgive you for your send last night. Will you agree to work with me to get back what we had before the kids came into our lives?"

Paul hugged her and held her tight. "Oh, God, yes!" Holding her away and looking into her eyes: "But I'm going to be self-conscious about it, at least for a while. Nate and Anita will catch our sends. We won't be able to prevent that."

"Nate has been watching us when we make love. Didn't you notice?"

"That little bastard! I'll . . ."

"No! Don't even think about it. Don't punish him for being curious. We need to talk to them about it, and explain that we will be sharing with each other as we used to. Please, Paul, I know enough psychology to know that we need to be positive about it when we talk to our children about sex."

"All right. Yeah. I know. You're right. I guess I'm a little prudish. . . ."

"A little?"

"OK, so maybe I'm a prude, but you love me anyway."

Bel laughed. "Very much." She sent love, surrounding him with it. She added an image of a bond between their hearts made of something stronger than steel.

Closing his eyes, he said, "Thank you." He sent his love, surrounding her and interpenetrating her love.

After a minute of bliss, she said, "Honey, we need to talk more about what's important to us. Do you want me to start?"

"I want to hear what's important to you, but I need to tell you more about George. OK?"

"Yes, you go first."

They walked slowly, arms around each other. Paul explained the importance of his invention and his decision to get credit for it. "Ralph's got my back, and he's going to make sure I get the lion's share of the credit. I had to write up some preliminary stuff about the interface so he can do that. I finished that before I came home. I need to do a lot more work on it, but he has enough to support me for a day or two. Now, do you understand why I was so upset?"

"Yes, honey. I do. You were pissed at George for what he did, and you were even more scared of what you might do in retaliation."

"Exactly. Now, please tell me how you feel."

"I'm still pissed at you. That send was nasty."

"Yes, it was. You have every right to be pissed. What can I do to make it up to you?"

"I'll figure something out."

"I bet you will." He hesitated and added, "But whatever it is, I'll do it. I would do anything to keep your love. I need you more than life. I already knew that, but Tony's death really brought it home to me."

"Good! I am sorry about Tony. I really am."

"Thanks, but the fact is that he's OK, and I'll be seeing him again. Now, please tell me more about what's important to you."

She took a deep breath, and explained about the stage-II colon cancer and the danger of a lawsuit. "Recent law states that a doctor is responsible for proper diagnosis and early treatment of cancer. With our current diagnostic techniques, there's really no excuse for a misdiagnosis. But the stupid guy didn't come in for his periodic exams. Oh, well. Enough about him. Today, he signed a release saying it was his fault. The other thing was a patient who has a dangerous transit. I referred her to a specialist, but he didn't do shit. Anyway, today I referred her to another doctor, and she'll be fine."

"You've said that influences from transits can be manipulated. Maybe some time you'd tell me more about that."

They were approaching the house. She said, "Yes, I will. Now that we're talking more, I'll be a blabbermouth and teach you to be a doctor."

He laughed. "Probably only enough to be dangerous."

As they entered the yard, Nate and Anita came running out.

Anita asked, “Is Pan OK? Why are you still carrying him?”

Paul said, “Oh, we just forgot. He’s fine.”

He laid the cloth carrying sack on the ground. When he loosed the end, Pan walked out with a strut, as if, “I meant to be in there.”

They all laughed except Pan, who ignored them as he rubbed against Anita’s leg.

I know this story was lengthy, but please allow me to explain. If I had listed all the principles, techniques and technology that I included in this story, the list would have been boring and uninformative. Instead, I told you a story that represents the potential impact on daily life of the scientific investigation and subsequent technology associated with nonphysical LORs. I do not claim that the details are correct, but I would like to make a few observations before we move on in our journey through time and thought.

First, I mentioned reincarnation. It implies that when a person dies, he or she may be conscious on other LORs and may visit loved ones. Furthermore, at some future time, the person may be born again, and have another full life. These are common beliefs among many people in the world, so I thought it best to include them. Actually, I had direct experience with death when my parents died, and a variety of experience that corroborates many of the common beliefs about reincarnation. Since the topic does not relate to fundamental research, I'm not describing it here. However, if true, reincarnation has many implications, including answers to the age-old question—what will happen when I die?

Second, I suggested that a composite machine might be able to use energy from the etheric LOR to move a physical vehicle. I think it likely, but research will be required to determine if it is possible.

We've already discussed medical uses of astrology, RV and related techniques and principles. I think the medical aspects of this story are quite plausible, based on our previous discussions. Remember Hippocrates' statement, "He who does not understand astrology is not a doctor but a fool."

As to the common use of telepathy and etheric awareness, who knows? If everyone grew up knowing that these are valid, what could we achieve? Scientists and technologists would attempt to improve on related training techniques, just as the RV people have. Parents would encourage their children to be more aware than other children, to give them an advantage in life. Teachers would further encourage the development of the children. I don't know where it might lead, but I think the uses of telepathy and other-LOR awareness are plausible as presented in the story.

And what if interstellar travel is possible by shifting matter to other LORs to avoid the special theory of relativity speed limit on the physical LOR? The potential impact on society would be tremendous.

I think we have covered enough for me to explain more about my perspective and my agenda. First, keep in mind that human perception is not the subject of this work. The subject is the reality in which we exist, of which we are only able to perceive a tiny bit. Our anomalous perceptions suggest that there is more than physical reality, but until now, it has been difficult to be sure. You will find that later chapters and appendices provide the means and methods with which we can investigate nonphysical LORs.

For the moment, consider two things that are easily verifiable: [1] science included some investigation of nonphysical LORs (e.g. astrology, alchemy and Hermeticism) until the Burning Times, and [2] the CIA research at the Stanford Research Institute provided strong scientific evidence of the validity of some aspects of nonphysical LORs. I have shown you other evidence, but you can verify these two facts with little effort. In combination, these two facts suggest that nonphysical LORs are real and have always existed.

To this, many people would respond, "OK, so some people have always been psychic, and psychic phenomena are real. So what?" That reaction places all the emphasis on the observers, who people commonly call "psychic." If someone tells you there's a diamond ring at your feet, do you stare at that person, or do you look down to see the ring? If someone tells you there's a tiger coming out of the jungle toward you, do you stare at that person, or do you look around for the tiger?

My point is that focusing on the observer prevents us from learning about what they have observed, e.g. nonphysical LORs. The point of this work is the investigation of nonphysical LORs—and my agenda is to promote that investigation.

We have been taught that there are three physical dimensions—left-right, forward-backward and up-down. Time is considered to be a fourth dimension. My perspective is that there are and have always been four dimensions plus time. Everything we know about, including ourselves, actually exists in all four dimensions progressing through time. In diagrams, we normally represent the three physical dimensions with x, y, and z, and use "t" to represent time. To these, I add "w" (as shown in Figure 1).

The levels in the w-dimension are continuous like the levels in the atmosphere, and like the atmosphere, certain ranges are given names. In the atmosphere, ranges of levels are commonly called the Troposphere, Stratosphere, Mesosphere and Ionosphere. Ranges of levels in the w-dimension are commonly called the Physical, Etheric, Lower Astral, Upper Astral, etc. These categorizations of levels are useful because significant differences exist between them.

In general, our physical senses have evolved to help us find food and to avoid being food for other predators. Thus, our eyes see wavelengths and our ears hear frequencies that are adequate for these purposes. In our evolution, it has not been necessary for us to see germs or to be consciously aware of non-physical levels in the w-dimension.

Never the less, humans, and all other creatures as we know them, evolved in four dimensions through time. That is my perspective. Now consider the implications if I am correct. If the human organism evolved in those four dimensions, not just the three physical dimensions, what forces influenced our evolution? What forces still influence us, based on that evolution? I recommend that you keep these questions in mind as we proceed on our journey.

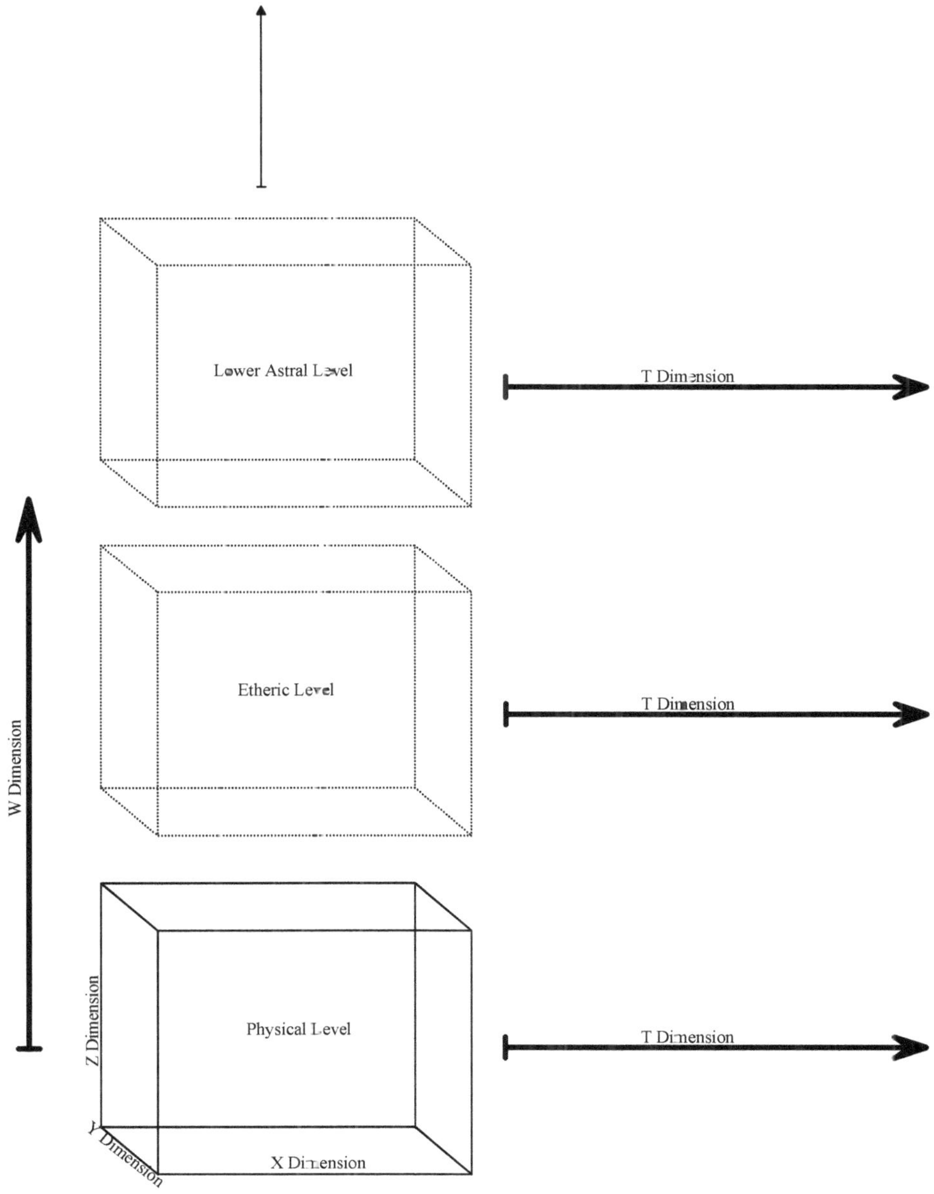

Figure 1 Dimensions

Section VI: Research Methods and Issues

Chapter 11: Experimental Design and Scientific Observation

On this leg of our journey, we will take a brief look at how scientific observations could be applied effectively in the realm of nonphysical LORs. I think all scientists would agree that objective and accurate observations are of central importance in performing scientific experiments. Unfortunately, perfect observations are an ideal that is not possible to achieve in any field of scientific investigation. All observations include some potential for experimental error. Perhaps the ultimate in this sense is the Heisenberg uncertainty principle, which states that we cannot, at any point in time, accurately measure both the position and the momentum of an object.

In this chapter, I wish to cover the types of potential errors an experimenter's observations may introduce in the context of nonphysical experiments, and how those errors can be minimized. I will not claim completeness in this discussion, but if I can heighten your awareness of potential problems and help you to improve the accuracy of your nonphysical observations, I will have succeeded.

First, let's consider the context of the observations that will be needed. At this time, I do not recommend further experiments to learn about the types and extent of human potentials. If you wish to learn more about human potentials, I suggest you study the experiments that have been run by parapsychologists over the last 60+ years, and the most recent discoveries of research with remote viewing. I think we need to learn more about basic principles and facts about the etheric and astral LORs, and that only then will we be able to understand the human potentials observed by parapsychologists and often labeled, "psychic."

I propose a set of basic experiments to support initial investigation of nonphysical LORs. These experiments chiefly fall into two categories. The first category consists of experiments to study astrological correlations. The second category consists of experiments to directly observe and learn about the etheric and astral LORs, primarily the etheric LOR, since that is the easiest LOR to observe, next to the physical.

Most astrology experiments involve studying astrological principles with physically observable effects, but the effects are largely observable only in human behavior and reactions—just as in psychology experiments. Thus, I suggest that the experimental methods and procedures can be copied from psychology. However, if you wish to be even more stringent in your procedures, you may borrow methodologies from some of the more carefully designed parapsychology experiments.

I recommend that astrology experiments be designed to include the following principles.

[1] Experimental choices would be double-blind.

[2] The subjects probably would not be given accurate prior knowledge of the expected behavior, and may be intentionally misled or misdirected.

[3] Each experiment would be designed to provide both positive and negative evidence related to the hypothesis being tested.

[4] Control groups would be part of the experimental design.

I will describe each of these principles in a little more detail, but books have been written about experimental design, and taught in college courses. If you are unfamiliar with the principles of experimental design in psychology, I suggest a college course or further reading.

To better discuss experimental design, let us consider a pattern that might be used for a wide variety of astrology experiments. First, there is the researcher. The researcher designs the experiment and coordinates its execution, often performing the data analysis, followed by publication. Next, there may be research assistants. These people execute the steps in the experiment and collect the data. Third, we have subjects. In an astrology experiment, the research assistants may observe the behavior of the subjects as they are affected by astrological influences, or the subjects may be reacting to the effects of influences they observe in others. In the latter case, we may call the influenced people, the "targets." You will see this pattern in the chapter titled A Simple Sun-sign Experiment, which is designed to illustrate the correlation between personalities and sun-signs.

In the context of this pattern, a double-blind approach keeps the research assistants in the dark with respect to the expected results. I.e. the people collecting the data have no idea which of the possible responses they should expect from a given subject. For example, in the case of the simple sun-sign experiment, the researcher will prepare the materials to show the subjects and make note of which selection is expected, based on the astronomical calculations. When showing the materials to the subjects, the research assistants will not be privy to the "correct" response until after they have collected the data. This makes it less likely that a research assistant might inadvertently give clues to a research subject.

In astrology experiments, the most unbiased results may be achieved by misleading the subjects. It is probably best to avoid even mentioning astrology to the research subjects. Some people know enough about astrology to bias the results—potentially. But if they think the experiment is a psychology study, or some other misdirection, the results are less likely to be biased.

The experiment should of course be designed to show direct evidence of the validity of the hypothesis or the lack of such evidence. But in addition to this, I suggest that you think about potential biases that might be real or imagined by critics. When possible, design the experiment to demonstrate the presence or absence of such biases. The simple sun-sign experiment does this, for example. The primary purpose of the collected data in that experiment is to determine if the subjects perceive the expected personality traits in the targets. However, what if the subjects somehow realized the "correct" personality traits based on conscious or unconscious knowledge of the target's sun-sign? If this were the case, then roughly the same rate of success would be expected for the targets that are excluded (see the experimental design for explanations of which targets are excluded). Thus, the same data is collected for the excluded targets, but analyzed separately. If the rate of correct attribution is significantly different, it suggests that attributions were not based on the subject's knowledge of the target's sun-sign. This is useful evidence whether or not critics bring up the issue. I also suggest that if a critic sees a potential flaw in your experiment, that you consider modifying the experiment to check.

Including a control group in an experiment is usually simple, although it increases the level of effort to perform the experiment. In the case of the simple sun-sign experiment, a control group can be a subset of the targets, for whom no correct attribution is offered. I.e. none of the three suggested personality descriptions matches the target's sun-sign, and one of the three is arbitrarily selected as being "correct" prior to showing them to the subjects. Thus, we would expect random results from the control group. Unfortunately, it would be easy for the researcher to make a mistake in setting up the control group and the "arbitrary" selections. In astrology, some of the sun-signs are more similar than others, and the more similar sun-signs tend to be numerically related to each other. Thus, I suggest that all randomization be done by computer, with no built-in biases. E.g. when selecting sun-signs for the control group, do not consistently select the fourth sign from the target's actual sun-sign. Simply program the computer to randomly select three of the eleven signs not matching the target's sun-sign.

I think that covers most of the potential issues with observations connected with astrology experiments. Data collection for these types of experiments depends on our normal physical senses, and is easily understood.

Direct investigation of nonphysical LORs is not as easy at our current stage of learning. I will attempt to cover some of the more basic issues in making direct observations of etheric and astral level phenomena and objects. You may have had training on making accurate observations, such as from a police force or from the FBI. If so, I hope that you will see similarities in my suggestions. Also, if you have had such training, you could probably add useful comments and techniques, as I have not had that training.

Here we go ...

First, I need to explain a little about the problem. Let's consider an experiment to measure and classify the effects of a nearby microwave source on a human aura. I hypothesize that under the right conditions, the aura is affected by changes in the intensity of a nearby microwave source even though there is no change in physical microwave radiation at the target aura. You will find this experiment described in a later chapter. I contend that the researcher should make direct observations in the course of this experiment. Research assistants may also be employed, at least some of whom are also capable of making direct observations. Such observations are made on the etheric LOR—not the physical LOR.

In my experience, only extremely sensitive psychics are able to make such observations with consistent clarity. On the other hand, most people are capable of learning to perceive the etheric LOR to a useful degree, so let's assume that the researcher has learned to perceive the etheric LOR—i.e. has developed a useful degree of etheric vision through training. In my experience, the etheric vision of such a person will initially be quite vague and dark. The degree of clarity will be different at different times, as with any learned ability. The clarity tends to improve over time, with the right types of practice. So the question is, what types of things make perception more difficult, and what can be done to improve consistency and clarity?

First—strong beliefs. These can prevent accurate observation. For example, if you believe that something is not possible, then you will not be able to see it, no matter how obvious it is to anyone else. On the other hand, if you believe you will see something, you probably will, whether it is there or not. When your perception is subtle or vague, even a weaker belief may make it difficult or impossible for you to perceive what others readily observe. If you have strong beliefs, you are unlikely to be objective in the areas controlled by those beliefs. As such, you should probably avoid attempting scientific observations where your beliefs might interfere with your perceptions.

In the context of perceptions, such as etheric vision that is as yet vague and indistinct, expectations are similar to beliefs in their effect. For example, if you expect to see the etheric equivalent of a leaf, but you are looking at a stone, you are likely to think you see a leaf. Only with a great deal of experience are you likely to recognize the stone and accept that your expectation had been incorrect. Unlike beliefs, you can adjust your expectations before you attempt to make an observation. It is best to explicitly set no expectations, being completely open to whatever might be there.

Strong emotions can also warp your perceptions. For example, if you feel deeply depressed or angry, you are less likely to recognize something that is peaceful or beautiful. You are more likely to perceive something dark or ominous. As such, I recommend that you make your scientific observations when you are feeling neutral—no significant emotions. Again, with more experience, the effect of emotions will become less significant, but even the most experienced psychic can be influenced by strong emotions, resulting in biased or outright incorrect observations.

Physical senses utilize a relatively high percentage of a person's brain, and are likely to influence your awareness of nonphysical LORs. Let's consider an example. Suppose you were watching a scene that was quite unfamiliar to you. For example, if you have never scuba-dived, imagine wearing scuba gear, complete with a mask and breathing through a hose from a tank on your back, eighty feet below the surface. The light is weak and wavering. You are watching fish in a coral reef, with crabs wandering about, with a moray eel attempting to catch fish. Now suppose you have a strong itch on your left shin, one eye hurts, and someone is shouting at you from an earphone in your right ear. How clearly do you think you would be able to observe the way the eel moved? In contrast, if you were watching the same scene on a big-screen high-def TV from your living room couch where you were completely comfortable and without distractions, would you be more able to follow the movements of the eel? Of course. Thus, it is best to minimize physical stimuli when attempting to make scientific observations.

If the environment eighty feet underwater sounded unfamiliar to you, consider the etheric LOR. If you are like most people, you have been conditioned to think that no such environment even exists. Thus, you may have no conscious experience there. The human mind requires experiential reference points in order to perceive and categorize. You have seen pictures of fish and crabs and moray eels, so those are all familiar concepts and have associated pictures in your mind. To illustrate the point, consider a chair. You know what a chair looks like. You have probably seen thousands of them, in various sizes and shapes. So when you see something that looks like one of the chairs you experienced in the past, you recognize it as such and make assumptions about its capabilities and purpose. In other words, you have ample experiential reference points to perceive a chair. However, you have no idea what you may or may not see on the etheric LOR. You have no effective reference points in your conscious experience. In fact, it's probably misleading to call it etheric "vision", since you might expect your impressions to be like your physical sight. Remember that your mind will attempt to attach your new experience to existing reference points. As such, if you were to perceive a human aura, you might sense an odor, see bright colors, and hear a sound. Such initial associations may be misleading. I suggest that you keep an open mind and compare notes with others. Again, experience is the best solution to this observational problem. While gaining experience in perceiving nonphysical LORs, I recommend that you test your observations frequently, so that you will not become accustomed to false interpretations. You will find more discussion of this principle in a later chapter about the development of etheric vision using the AWIN system.

Mental or physical fatigue reduces your ability to make accurate and unbiased observations. I suppose this is common sense, and requires no additional comments, but if you question this statement, consider that the brain is involved in interpreting your observations. If you are sufficiently tired, the brain is not capable of functioning at its normal level of intelligence. Thus, rest may improve your scientific observations.

A full stomach is another negative factor in etheric vision. Significant hunger is a similar problem, causing a type of pain (perceived as hunger) and also fatigue. Again, with sufficient experience, the reduction in perceptions is minimized, but it is always best to make scientific observations while you are neither full nor famished.

This one may fall under the previously listed categories—a feeling of hurrying tends to increase tension, and reduce the accuracy of your perceptions. Thus, I recommend planning your observations so that you will not feel like you have to hurry.

To summarize, I have listed a number of potential difficulties that you could avoid or work on before attempting any serious observations of the etheric LOR. Allow me to turn them around and state them here in the positive. To maximize your ability to make objective and accurate scientific observations of the etheric LOR, I recommend that you

[1] Gain at least some experience of the etheric LOR and verify your perceptions by comparing notes with others who are developing, or have developed, their own etheric vision,

[2] Make observations only in areas where you have no strong beliefs, and make note of any beliefs that may relate to your observations,

[3] Use self-talk or other mechanisms to set no expectation or to set a mild expectation that you will observe something new or unexpected, and reassuring yourself that you have plenty of time,

[4] Be calm, rested and in no hurry, having eaten enough, but not too much or within the last hour,

[5] Arrange your observations so that your (closed) eyes will have darkness or constant low lighting, your ears will hear nothing, your nose will smell no odors, and your body will be comfortable.

I have also found that perceptions can be enhanced by making observations in a group. I.e. if you have several friends who have etheric vision or are developing it, and you make an observation together, your perceptions may be sharper and more detailed. Just be careful not to allow any hints or discussion that might set expectations. For example, suppose you have two friends who wish to participate in an experiment to observe the effects of a nearby microwave source on your dog's aura. To utilize the group-effect, you would first set up the experiment as if you were the only observer. No discussion would take place among you and your friends about what you may or may not observe prior to or during the experiment. After the experiment is completed, the three of you would make separate detailed notes in silence. When all of you finished journaling the experience, you would discuss it. You would probably find the discussion reminds you of impressions you received during the experiment that you had forgotten to note in your journal, and you could add such notes after or during the discussion.

You would probably also find that this phenomenon of enhanced group observation is analogous to a physical perception test, as follows. Suppose that you and your two friends walked into an unfamiliar room together, looked around, then walked back out without a word and took individual notes about the room. You would find that each person's observations, as reflected in their notes, were different, but in a subsequent discussion of the room, the combination of your three perspectives would develop a more complete and more accurate description of the room than any of you could produce separately. Thus, in addition to sharper and more detailed perceptions, the combination of your observations also improves your composite understanding of what happened during the experiment.

All right, so we've talked about some principles for making useful scientific observations in experiments on nonphysical reality. But if you don't already know how to perceive other LORs, how can you begin? Which training system should you use? What can you hope to accomplish? I'll try to answer these questions in the next few chapters.

Chapter 12: Remote Viewing

This leg of our journey will be interesting, but brief. Several systems teach the development and use of Remote Viewing, or RV as it is often called. I have looked at RV training, but have not taken it. The techniques and principles seem sound to me, and the reported results are outstanding. For the purposes of making direct observations in scientific experiments related to nonphysical LORs, I think that RV is an excellent choice. Qualified instructors are available in most areas of North America, and complete courses are available on DVD.

One down-side of RV is that you may need to modify the techniques slightly, since RV is most often used to view remote physical locations or events. However, I'm sure that if the system you choose does require modified techniques, the modifications would be minimal.

RV techniques generally take the issues I listed in the previous chapter into account, avoiding many of the potential pitfalls simply by the methodology. Some of those techniques are specifically designed to get your conscious mind out of the way. I would expect this to minimize the effect of beliefs, expectations, etc. This has its advantages, but also its limitations.

Another plus for RV is the emphasis on verification. I.e. you are encouraged to check your perceptions against reality. For example, if you were attempting to use RV to look at a particular intersection in a city, you would be encouraged to check your RV session results against pictures of that intersection.

Some of the RV systems claim that you can learn to have useful RV sessions in about 30 days. Naturally, that is only the starting point, as practice and further training are necessary to improve the accuracy and consistency of your observations. On the other hand, you might be able to begin nonphysical LOR research after only one month of RV training.

Lacking experience with any of the RV systems, I hesitate to suggest one of them in particular. However, if I were to choose one, I would look for one that has benefited from the CIA and Army training programs (derived from the work at Stanford Research Institute). Thus, I would pay attention to any recommendations from Russell Targ, H. E. Puthoff, or Major Ed Dames.

I suppose you may be asking why I haven't delved more deeply into RV. I think my reasons will become clear in the next few chapters.

Chapter 13: Etheric Awareness through AWIN

On this leg of our journey, we'll go into the etheric LOR. Well, at least we'll talk about a training system that is designed to assist in the development of controlled awareness of the etheric LOR. This is a training system that I have developed over the last 40+ years. I call it the AWareness and INtegration (AWIN) system. [80] AWIN is based on the Qabalah, which is the Jewish mystical tradition. Dion Fortune was a well-respected author and a member of the Hermetic Order of the Golden Dawn. The Golden Dawn was formed in the mid-19th century, a few years after the Inquisition lost its death penalty. According to her, the Qabalah dates back to the days of the Old Testament, [81] but until the 15th century it was passed down in verbal form with glyphs such as the Tree of Life. However, other Qabalistic documents, such as the Sepher Yetzirah, predate the 15th century. The Qabalah is also known as the "Yoga of the West."

But I have no deep interest in the age or origin of the Qabalah. The Qabalah is a practical system for understanding and experiencing the reality in which we exist, and that *is* my interest in the Qabalah. In a very real sense, the Qabalah is like a drivers' manual combined with a travel guide. You do not need to believe in cars to drive one. Neither do you need to believe in anything to use the Qabalah to experience the various LORs. But you do have to learn to drive.

---

[80] http://www.awinsystem.org/index.htm

[81] Fortune, Dion. (1972) *The Mystical Qabalah* (p. 3). London, UK: Ernest Benn Limited, Tenth Impression.

In this chapter, instructions are provided to teach how to perceive the etheric LOR. This is analogous to learning to see out the windshield of a car. The next chapter deals with astral travel, which is analogous to actually learning to drive a car. The "travel guide" is not covered in the present work.

As with driving a car, it is easier when you have a qualified instructor. However, many people are able to learn to use the AWIN system without a teacher. If you seek a teacher, make sure you select someone who is able to perceive the etheric LOR. Would you want to learn to drive from a blind person? It's the same principle. A person must be able to observe what you do—or fail to do—in order to teach you how to do it. If you join an organization to find a teacher, make sure that the organization will allow you to publish your research.

You do not need to study the Qabalah to learn etheric vision and astral travel. In fact, I do not recommend it for these purposes, as some people get so excited that they abandon basic research and dive deep—as I did. Even though I recommend that you initially avoid learning about the Qabalah, I wish to state a few opinions. First, different schools/organizations use different approaches in teaching the Qabalah. The Kabbala Center apparently teaches a simplified version that makes the knowledge accessible, to some extent, to more people—which I do not recommend to anyone who might be serious about other LORs. Some Jewish organizations teach a more traditional version based on a complete understanding of the Zohar. It is my understanding that some of these organizations focus more on an intellectual understanding than on an experiential basis. I suggest that an experiential basis is crucial. Do you want to be able to drive to interesting places, or will you be satisfied just talking about driving? (My opinion. Just my opinion.)

The Golden Dawn teaches a modern and comprehensive system. Before joining such an organization, find out how much experiential (i.e. practical) work they do in the classes. When I teach the AWIN system, I try to include some practical exercises in each class session.

The AWIN system begins by teaching objective awareness of the etheric LOR. Since this is precisely the type of perception needed for fundamental research in the etheric LOR, I will describe how to develop and use this basic mode of perception. With consistent weekly effort, you should expect to learn basic etheric vision in about 30 days. If your efforts are less or less frequent, you can expect your ability to perceive the etheric LOR will accordingly be less or take longer to develop.

With the AWIN system, you can develop awareness of other LORs as an addition to your physical senses, in such a way that your conscious mind can continue to work actively just as it does when hearing a conversation or smelling food. But you are unlikely to be able to accomplish this in one or two months. Initially, you will need to use mental silence to enhance your perceptions. With sufficient practice, you will learn to continue to observe etheric objects and events while your mind analyzes your observations. The following may be helpful in understanding this aspect of human psychology.

If you had a "normal" childhood, your parents taught you to ignore your perceptions of the etheric LOR, while encouraging you to develop accurate perceptions of the physical LOR. You learned to see with your eyes, hear with your ears, feel with your hands, smell with your nose, and taste with your tongue. If your parents had taught you that your hearing was just your imagination, today you would not be able to talk and you would not know how to interpret what you hear. Effectively, you would not have developed your hearing—just as you have not developed your ability to perceive the etheric LOR. Your mind has been trained to ignore such perceptions.

This is why your mind must be silent while you are developing your ability to perceive other LORs. Your mind has developed the *habit* that such perceptions are only allowed while you are sleeping. Now you must teach your mind that it is OK and appropriate to be aware of the etheric LOR while you are awake. This requires the development of new habits, which is what this training is all about. I.e. think of this training as the development of new mental habit patterns. These new habits must be strong and consistent if you wish your perceptions to be clear and consistent.

Before we get into the detailed principles and techniques, I would like to make a few comments to set your expectations. First, do not expect the etheric LOR to be like the physical LOR. What you will perceive on the etheric LOR will be different, although there are similarities. In order to learn to perceive the etheric LOR, you will need to use what you know—from experience with the physical LOR. Do not let that confuse you. For example, you will probably see light on the etheric LOR. It is not electromagnetic radiation, meaning physical light. You will also see etheric objects. Etheric objects are not made of physical matter, although they may coincide with physical objects that have the same or at least similar appearance to your physical eyes. However, you can also perceive physical objects from the etheric level. OK, so it can be confusing. My point is that you should not make assumptions about etheric objects and events based on your experience with the physical LOR.

Second, do not expect your perceptions to be as detailed as your physical eyesight on a bright day. If you set the expectation of detailed perception, your mind will tend to manufacture and fill in the details. For the purposes of scientific observation, you want to perceive exactly what is there, with nothing added and nothing taken away. Therefore, I suggest that you set the mild expectation that your perceptions will be vague. Whenever possible, verify any details that you do perceive.

Third, the etheric LOR is usually perceived as being dark, as on a moonlit night. There are exceptions, but this is usually the case.

Fourth, the etheric LOR is usually perceived as quite viscous, at least by less experienced people. This will not affect your etheric vision, but if (when) you learn to enter and use your etheric body, you will probably find that it is difficult to move.

Fifth, why do most women dream in color while most men dream in black and white? I don't know the reason, but I've found that the same is true for etheric vision. I.e. most women see the etheric in color while most men see it in black and white—most of the time. For myself, I find that I usually see the etheric in black and white, with occasional splashes of color. Maybe it's related to the structure of the human eye. The eye has rods and cones. The rods register only black and white. The cones in the eye register colors. Maybe the analogy doesn't work, but it's interesting. In any case, don't be surprised if your etheric vision has similar color (or lack thereof) to your dreams.

OK, on to the use of AWIN to develop etheric vision. If feasible, form a group to learn etheric vision together. The ideal group size is between five and twelve. I suggest no one under age thirteen, and the participants must get along reasonably well with each other. Some people will learn the techniques more easily than others, so you may notice that one member may be immediately successful while another member struggles for months to perceive an aura. With at least five members, you will probably be able to give each other useful verification after journaling each session.

Begin with Meditation Method 1 [82] which is an MP3 file. You can play it directly or you can copy it to an MP3 player. Before playing the MP3 file, read about the use of the meditation methods in Appendix D. Then follow the directions to learn Meditation Method 1. You should practice this first method for at least a week and at least three times. For example, a good schedule would be Monday, Wednesday and Friday. While it is OK to practice these methods several times per day, practicing three times on one day is not as effective as practicing on three separate days.

It is especially important to be undisturbed while you use these meditation methods. Turn off your phones. Tell everyone not to disturb you. Do not answer the door. Do not allow your pets to come near. Remember that you are developing new mental habits, and you want those habits to be effective. If you come out of levels without following the prescribed methods, meditation will become more difficult for you and you will be less likely to make objective observations of the etheric LOR. If you are jarred out of levels once, don't worry about it. If it happens twice, your meditation will probably be less effective. If it happens often, there is no point in continuing, and you might as well leave the research to other people. I highly recommend journaling each experience, meaning each time you practice meditation. I suggest including the following in each journal entry.

---

[82] http://www.awinsystem.org/Handouts/MedMethod1.mp3

[1] ES: Emotional state

[2] MS: Mental state

[3] TSA: Time since last ate

[4] FS: Fullness of stomach

[5] R/A: Degree of restedness/alertness

[6] AIP: Level of pain

[7] DIL: Degree of illness/health

For convenience, I suggest that you format your journal entries with these or similar labels to simplify your note taking. The abbreviations are for your convenience. One purpose of the journaling is to learn which conditions (circumstances) affect your effectiveness in meditation, and especially in your etheric perceptions. If you think some other factors may affect you for good or for ill, keep track of those as well. The body of your journal entries should of course describe what you experienced, including the level of clarity, details, and verification, if feasible.

Repeat this practice for Meditation Method 2. [83] I.e. follow the directions, practicing the method for at least a week and at least three times. Then repeat this for Meditation Method 3. [84]

Each successive method is shorter than the previous one. What you are doing with these methods is training yourself to relax physically, mentally and

---

[83] http://www.awinsystem.org/Handouts/MedMethod2.mp3
[84] http://www.awinsystem.org/Handouts/MedMethod3.mp3

emotionally, while becoming aware of your aura and etheric body as separate from your physical body. With sufficient practice, you will see that you can move your aura and etheric body partially independently of your physical body. In fact, this can lead to astral travel, but that's the topic of the next chapter.

When you have learned Method 3, you are ready for additional exercises. You can then practice meditation without the use of the MP3 files. You will easily remember the relaxation sequence and you can follow it without prompting. Here is some simple and unambiguous terminology. Entering meditation is called "going down in levels" while ending a meditation is called "coming out of levels." In this context, the word levels refers to levels of relaxation and not to LORs.

Your first exercise is given in Appendix E. I call it the Comet Exercise for reasons that will probably become obvious when you do it successfully. Follow the directions. Don't get creative about it. Keep the ball of energy inside your physical body. Move the ball at a consistent rate. Gather the energy slowly. Release the energy slowly. It is not important that you return the energy exactly from whence it came, but it is important that you don't gather too much from any one area or release the energy in any one area. For the time being, it is important that you take this on faith. When you gain more experience, you may be able to see the centers in your aura that are typically called "chakras." If not, you may wish to experiment and see what happens when you mess with them, but be careful. It may be possible to injure yourself by moving too much energy into or out of a single location in your aura.

With the Comet Exercise, you can learn both perception and control in the context of your aura. This is a starting point for your etheric vision. If feasible, get verification from someone who has effective etheric vision. I.e. the Comet Exercise is not intended to be a mental or imaginary exercise. If you are actually moving energy in your aura, that is an objective activity that others can observe, just as someone may observe how you drive a car.

You should also practice the other two exercises in Appendix E. The Human Aura Exercise is most effective when you are working with multiple members who are developing together. As you can see from the exercise, one member moves (changes) his (her) aura while others observe. The Plant Aura Exercise can be used when you are learning these techniques without a group or class.

If you are learning in a group, you can think up many additional exercises to further develop your etheric vision. If you have a teacher, he or she should already have appropriate exercises for you.

I suggest using Meditation Method 3 for many months, until the process of entering meditation seems slow and redundant. Then I suggest that you transition to something like Method 4. By this time, you should have sufficient control over your etheric vision that you can perceive the etheric, at least to some extent, simply by shifting your focus.

I think some explanation is in order. We've talked about the what and how of this training. We've even talked a little about what you can expect to experience. But we have not talked enough about your experience while learning the new mental habit patterns needed for etheric vision. Keep in mind that you do not initially have experiential "pegs" for your etheric vision. Your ability to perceive will develop as you gain those memory pegs, and the more accurately you can interpret your perceptions, the better will be your progress. For example, imagine that initially your eyes are closed. You have never used your eyes, but someone gives you a technique with which you can open your eyes a little at a time. (E.g. you had an operation that allows you to use your eyes which have been dormant until you were an adult.) At first, your vision is hazy at best, and your brain has to develop ways of interpreting this new input while at the same time trying to understand how to control your eyelids and eyes. You think you perceive something, so you reach out with your hand and touch it. Suppose it is another person. That person can tell you that you are touching him or her. Now, you have a point of reference—you have perceived a person, and you received verification. Your eyes were barely open and out of focus, but you know you perceived another person. With much more experience, you learn to open your eyes to let in enough light, you learn to focus your eyes, and you develop the ability to differentiate between different people by sight.

Etheric vision is developed in a similar manner. This process of development occurs at different rates in different people, depending on many factors, the most important of which is the amount of practice, with verification, analogous to the person seeing another person and touching them, then being told that they did touch the other person.

Suppose the technique that you were given to open your eyes initially was to pull down on your cheeks with your fingers. Meditation is used like that to help you to "see" on the etheric level of reality. In the analogy, as the person practices, her cheek and forehead muscles develop strength and she is able to open her eyes without using her fingers. Similarly, meditation becomes less important in perceiving the etheric after lots of practice.

Just as with physical eyesight, initially you have no reference points for what you perceive on the etheric. You need to practice as often as possible perceiving things that are verifiable. Although not perfect, probably the best approach is to work with someone who has already developed etheric vision. Let's call this person your "guide." Your guide should be able to watch what you do on the etheric LOR and perceive what you are attempting to perceive there. In a way, your guide is acting like that other person in the analogy about physical eyesight, giving you feedback.

I know that some of this has been repetitive, but I think some of the ideas bear repeating. Most people think of etheric vision as "imaginary" or "subjective." In my experience, it isn't. Many times, I have asked students to do something, watched what they did, and then commented on their actions. Initially, they have been amazed that my comments related to things that they thought only happened in their minds. I.e. they thought their actions were imaginary or subjective. In fact, I was able to observe those actions and give them feedback—like a teacher in drivers' training. This is precisely why fundamental research in nonphysical LORs is possible and desirable. Objective observations can and should be made to learn more about the nature of the reality in which we exist, which we influence and by which we are influenced.

Etheric vision is sufficient to make many types of direct scientific observations. Later chapters include many suggested experiments where such observations are needed. But etheric vision is only useful when you are in immediate proximity to that which you wish to observe. If you wish to go further, you should consider developing RV or astral travel (also known as astral projection).

Chapter 14: Astral Travel through AWIN

Astral travel is similar to Remote Viewing, but with a bit more flexibility. It is a little like flying a plane. If you stay near the physical LOR as you travel, you can actually navigate using a compass heading and landmarks, such as cities, mountains, rivers, etc.

There are two distinctly different approaches to astral travel. Describing these will be somewhat difficult, and even the basic concepts may be difficult for you until you gain some experience. One method utilizes your etheric body. What is your etheric body? Think of it as an object on the etheric LOR that usually looks like your physical body, and whose location usually coincides with your physical body. It is possible to focus your attention in your etheric body and move around in it—like you would drive a car. This is one type of astral traveling, and I recommend it for our current purposes, i.e. for the purposes of making objective observations in scientific experiments.

The second approach is to travel without your etheric body. I'll explain this approach in more detail below.

Traveling in your etheric body is similar to flying an airplane that has equipment that is more sensitive than your eyes. By using the equipment of the etheric body, you can perceive the etheric and physical LORs in more detail and with more clarity. But before going into more details about *what* you can do, let's first talk about *how* you can learn to do it. And it would be incorrect to infer from what I said that the etheric body has instrumentation analogous to your eyes. Although I have suspicions, I do not know why more clarity is possible by using the etheric body. This is one of the things I hope we can discover.

To learn either approach to astral travel, begin by developing significant experience with the etheric vision techniques as described in the last chapter. If you have a teacher, I suggest at least two months of experience, practicing the exercises and techniques at least three times per week, with regular coaching and feedback from your teacher. If you do not have a teacher, you should plan on twice this amount of experience before attempting astral travel. The following instructions assume you do not have an instructor. If you have an instructor, he or she probably has methods to teach you astral traveling. However, a teacher may also ask you to follow these steps. If so, your teacher can give you feedback at each stage.

ASTRAL TRAVEL IN YOUR ETHERIC BODY

The same type of meditation is used for astral travel as for etheric vision. Go down in levels (enter meditation). Become aware of your aura and focus your attention on the etheric LOR, as you have in previous exercises. Now imagine that your etheric body is vibrating at a low C, as played on a piano. Feel the vibration. Hear that tone. Feel your etheric body. Lift your etheric arm. (Your physical arm should not move at all, remaining completely relaxed.)

When you are successful in lifting your etheric arm without tensing any physical muscles, stand in your etheric body. After each movement in your etheric body, check your vibration rate; feel the vibration of the low C in your etheric body and hear that tone. Sit back down, so that your etheric body once again coincides with your physical body.

NOTE: Always move your etheric body back into you physical body in a deliberate manner. Do not *jump* back into your physical body and do not simply transfer your focus back to your physical body. The first may damage one of your bodies. The second will reduce your control. Remember that you are developing a new habit pattern. If you mess it up, the habit—your ability to sense and use etheric objects—will probably be less effective.

You should practice until you are able to stand in your etheric body, keeping your physical body completely relaxed, while maintaining the low C vibration and tone.

Most people are able to look around on the etheric after standing. If you have difficulty seeing in your etheric body, feel yourself opening your eyes. For example, after standing up in your etheric body, imagine opening your eyes and explicitly looking around.

When you are successful standing (while maintaining the low C tone) and looking around, move around the room in your etheric body. As you move, check the low-C tone and continue to look around. You should practice this until it feels normal and natural.

Remember that you shouldn't expect your perception of the etheric LOR to be clear and crisp. If you find that it is dark and vague, just keep working on it. Even if it remains dark and vague most of the time, you will find that it can be effective, especially when combined with the perceptions in a group/team.

Now that you have practiced moving around in a room, practice outside. Stand and move around as before. Look at the auras of trees or shrubs. Jump and fly. Practice jumping until you get used to the lack of gravity. You will find that you can move up and down as easily as you can walk around. Always check that you maintain the same tone, and verify your perception as often as feasible. If you find this difficult, imagine what it must be like for a baby. An infant takes many months to learn to see and walk, even with the support of experienced adults.

The tone that you feel and hear is an aid to maintain your focus on a specific LOR. Lower tones (such as the low C) will help you focus on lower LORs. You can use higher tones to help focus on higher LORs.

To move up to the lower astral LOR, shift the tone to a somewhat higher frequency, perhaps middle C or a little higher frequency. Yes, I have found that your etheric body is capable of moving on the lower astral. I worked with one person who was even able to rise into the upper astral in her etheric body. Experiment with the tones until you are comfortable moving up and down the LORs. As usual, verify your perceptions as often as feasible.

ASTRAL TRAVEL WITHOUT YOUR ETHERIC BODY

This approach is not quite as easy to describe as using your etheric body. Effectively, you should forget about having a body, and simply focus and move your attention. If you have a teacher, he or she can help "pull" you out of your etheric body. When traveling without your etheric body it is a little more difficult to remain focused, so practice focusing. Essentially, you should keep yourself compact when you leave your physical body. Use the same tone as you would use if traveling in your etheric body to focus at a specific LOR.

Practice is similar, except that you needn't practice outside at any stage. Movement is easy through all walls, doors and windows. With this method, you have no arms or body, per se. You begin with meditation. When you are ready, move forward out of your body. This is "simply" a matter of moving your focus forward. From there, check your tone and look around. Then move your focus back into your body. When you can do this comfortably, move farther away from your body. Always keep track of your tone. You will find it is easy to move in any direction.

NOTE: Again, always move back to your physical body in a deliberate manner. The reasons are the same as in the note for the use of your etheric body.

As I said above, I recommend using your etheric body for astral travel. My primary reason is that you will probably find your perceptions will be clearer from your etheric body. However, I have read that traveling long distances in your etheric body is more dangerous, especially over water. As such, I suggest only short trips in your etheric body. If you wish to make long trips, such as around the world or farther, I recommend that you leave your etheric body at home.

ETHERIC SCIENTIFIC OBSERVATION

Always remember that your perceptions are questionable. Even physical perceptions are frequently misinterpreted, and how many years have you been using your physical senses? Although you cannot completely trust your perceptions, you have nothing better to go on, so you need to develop your understanding of the etheric level of reality based on your experience and the relevant experience of others. Our scientific knowledge of physical reality has taken centuries to develop, with many thousands of scientists working on it. Our scientific knowledge of etheric reality must evolve in a similar manner, even if we take only decades instead of centuries.

Even after an etheric level phenomenon has been verified thousands of times by hundreds of scientists, our conclusions may change, just as our understanding of motion changed with relativity theory. And what should we try to understand? What directions could our initial investigation take most fruitfully? I have some ideas ...

Chapter 15: Fundamental Research Goals

On this leg of our journey, I'll suggest some principles for selecting the direction of fundamental research. I know that other scientists will have their own ideas about this, and my thinking is no better than many other scientists. I.e. I don't think this is the only reasonable approach. With that caveat, my thinking is roughly as follows.

First, I would like to consider some of the previous attempts to investigate nonphysical LORs. I do not wish to criticize other researchers, but I feel the need to make the observation that they have had limited success, and we should always learn from mistakes. In fact, I prefer to learn from other people's mistakes rather than repeat those mistakes. In general, I think it is safe to say that other researchers have investigated phenomena that they could observe to some extent. That is good in principle, but if the ability to observe is too limited, the observed phenomena may be misconstrued. For example, Carl Jung observed the correlations specified by astrology between natal charts and psychological patterns, but he could not perceive the influences. Thus, he wrote, "The fact that astrology nevertheless yields valid results proves that it is not the apparent positions of the stars which work, but rather the times which are measured or determined by arbitrarily named stellar positions." [85] In the absence of an observable causal relationship, Jung concluded that there *was* no causal relationship and explained the correlations as "Synchronicity." Observations are fundamental to scientific discovery.

---

[85] Roderick, Main. (1998) *Jung on Synchronicity and the Paranormal* (p. 13). Princeton, NJ: Princeton University Press.

In many other cases, researchers have focused on human potentials. To a large extent, experiments in human potentials are easier to replicate than any experiment that requires that the researcher have specific capabilities, e.g. the ability to make direct observations on the etheric LOR. Subjects are usually available who have various potentials, so a good researcher can replicate such experiments. My concern is that such research does not help us much to understand the nature of nonphysical LORs, including the basis for those human potentials For example, often-repeated experiments in parapsychology have demonstrated that some people are able to react to future events or circumstances. What is it that perceives those events or circumstances? We don't know. We haven't categorized (and verified) the nonphysical portions of humans in the context of these experiments. What is it that responds to that perception? We don't know. Some people may know the answers to these questions, but that knowledge is not wide-spread.

Assuming that you agree that direct observations are critical and that a different approach is needed, how should we proceed? First, we could look at documents from previous researchers who have looked at nonphysical LORs to discover the fundamentals. One such book was written by C. W. Leadbeater in 1895—*The Astral Plane: Its Scenery, Inhabitants, and Phenomena.* [86] Unfortunately, many such books have a religious slant, and are statements of fact (whether true or false) instead of providing a more scientific basis. Furthermore, I think that such books can only be used as

---

[86] Leadbeater, C. W. (1895). *The Astral Plane: Its Scenery, Inhabitants and Phenomena*. London, UK: The Theosophical Publishing House. Can be viewed online
http://blavatskyarchives.com/theosophypdfs/leadbeater_the_astral_plane_its_scenery_inhabitants_1895.pdf

reference, and we should design our own experiments and publish our results, just as in any other field of science.

Now, here are some of my ideas about how to investigate nonphysical LORs.

WHAT

The first thing to learn is *what* is there. I suggest that we categorize the *things* that exist in the etheric LOR. Until we know better, let's call them objects and energy. I've already suggested that you can find etheric bodies and auras on the etheric LOR. Those are examples of etheric objects. I've also suggested that you will find energy. Are there different types of energy? I think we need a scientific categorization of common objects and energy on the etheric LOR, a categorization that can be shared with and verified by other scientists.

RELATIONSHIPS

When we have a good idea of *what* exists on the etheric LOR, I suggest that we need to characterize the relationships among those objects and energy. I.e. how do they interact, and what effects do they have on each other. But we need to keep in mind that some interactions may occur between the etheric and other LORs. E.g. I have suggested that a physical microwave source affects some etheric objects. At this stage of our investigation we cannot quantify these relationships, but I think we can observe and make statements about causal relationships.

MEASUREMENT

In order to quantify objects, energy and relationships, we need to define measurements, i.e. units and the means to standardize measurement of objects and energy in terms of those units. Long ago, the height of a horse was measured in "hands." Distance was measured in strides. Power was measured in "horse power." These units were based on obvious convenient measurement methods, but they were not very useful for science until they were standardized and precise. The metric system provides this type of scientific standardization for physical measurements. My guess is that we may need to establish some rough measurements before we can become precise, but I could be wrong about that. For example, suppose we determine that an aura has a characteristic that we can call strength. How can we quantify the strength of an aura? We would probably define the units of strength in terms of the method of measurement. Suppose, for example, that we could measure the deformation of an aura by a certain intensity of a microwave source. Then we might define the units of aura strength as centimeters per watt at a distance of two meters from the microwave source. I know this is probably a naïve suggestion, but I think it makes the point that initial units may be defined by the method of measurement.

MATHEMATICAL RELATIONSHIPS

Having categorized etheric objects and energy, characterized the relationships among them, and defined units and methods of measurement, I think we would be ready to further delineate the relationships based on mathematical predictions. At this stage in our discovery process, we can make specific predictions. For example, if we were talking about astrological influences, we could predict a specific level of response on the physical LOR to a specific intensity and type of influence. The prediction might not be accurate, but having made the prediction, it can be tested, which is an important aspect of the scientific method—hypotheses with repeatable experiments, checked and re-checked by multiple scientists.

SUMMARY

We design experiments based on our ability to make observations, and I think that with RV and AWIN methods and techniques, we have the means to investigate the etheric LOR in more detail. In general, I think at this early stage we can be most productive by looking at simpler, more "atomic" objects, energy and events. Hopefully, such discoveries can be used as building blocks, like elements are in chemistry, so that we may later discover the nature of complex derived principles such as related human potentials.

Chapter 16: Balancing Skepticism and Openness

Along our journey, I have often mentioned the need to maintain skepticism and simultaneously be open to new possibilities. Carl Sagan wrote, "The scientific way of thinking is at once imaginative and disciplined. This is central to its success. Science invites us to let the facts in, even when they don't conform to our preconceptions. It counsels us to carry alternative hypotheses in our heads and see which best fit the facts. It urges on us a delicate balance between no-holds-barred openness to new ideas, however heretical, and the most rigorous skeptical scrutiny of everything—new ideas and established wisdom."[87] This is one of the most difficult aspects of the scientific method, and I want to say a few things about it from my perspective. If you are a researcher, I'm sure that you know these things just as well or better than I do, and I hope that you do not find my emphasis too annoying.

First, a researcher who is only skeptical cannot make progress. The only things worth investigating are those that are already known to be true. In Sagan's terminology, the facts don't fit the preconceptions and they are ignored. There is probably no researcher who is that skeptical. However, an overly skeptical researcher is less likely to consider evidence when that evidence contradicts, or simply does not support, "known" theories. As Sagan said, "Scientists do not seek to impose their needs and wants on Nature, but instead humbly interrogate Nature and take seriously what they

---

[87] Sagan, Carl. (1996) *The Demon-Haunted World: Science as a Candle in the Dark* (p. 27). New York, NY: Random House.

find."[88] Thus, a researcher must accept whatever they find, meaning they must be open.

But, a researcher who is only open and not skeptical can make no more progress than one who is only skeptical. The world is full of contradictory information, much of which is false. In my opinion, a researcher who is not skeptical will come up with sequences of random hypotheses that lead nowhere.

I wanted to emphasize the need for balance, primarily for the sake of those who are not scientists. Many people who study nonphysical LORs are open without skepticism. I've heard them say, "Anything is possible." Many people who do not investigate nonphysical LORs see no need to be open to the possibility. I've heard them say, "It's obviously wrong. Why should I waste my time on it?" Unfortunately, I think many of the statements made by those who are too open and not skeptical, *are* a waste of time. I suggest that balance is the way forward.

Here are a few more comments for less experienced researchers. First, beliefs and expectations of any kind are detrimental if they relate to the current experiment in any way. Even if the experiment has been performed a thousand times before, always with the same outcome, the next time it may turn out different. Be watchful for other possibilities. More importantly, if you hold beliefs about the outcome of the current experiment, it may influence your ability to accurately observe, record and analyze the results.

---

[88] Sagan, Carl. (1996) *The Demon-Haunted World: Science as a Candle in the Dark* (p. 32). New York, NY: Random House.

If equipment can take relevant measurements, it may help to reduce bias. E.g. the use of Kirlian equipment in combination with your personal observations might increase objectivity. Another excellent approach is to enlist the aid of others who have had the relevant training and developed the skills to observe the phenomena simultaneously. If a group observation is made, I recommend that there be no discussion prior to recording the individual observations. Only after each participant has recorded a detailed description of his/her observations can a group discussion be held without the danger of biasing the results. These independent journal entries provide some degree of balance, if the participants have conflicting expectations. Needless to say, if they all have the same bias, independent journal entries may not help create an overall objective result.

No matter how hard you may try to develop and maintain a balance between openness and skepticism, that balance will sometimes fail you. We all make mistakes. You will want to check and double-check your results. Then publish your results. One of the strengths of the scientific method is based on the replication of experiments, as every scientist knows. When you publish your results, other researchers can run the same or similar experiments, thus supporting or contradicting your results. If the same experiment is performed by 100 people in 50 locations, some of those people will be more skeptical while others will be more open. The result is most likely to be accurate because of this replication.

OK. Enough said about that.

Section VII: Research Directions

This section covers a variety of suggested directions for fundamental research into nonphysical LORs. Since the etheric LOR is the closest to the physical LOR, most of these suggestions will be on that level. Another reason for focusing initially on the etheric LOR is that the physical and etheric LORs are more likely to influence each other directly.

As previously stated, my suggestions revolve around the idea that we lack basic information about the etheric LOR, and that we need to develop that knowledge. As such, I suggest that the research proceed in phases. The first phase is to establish that something exists. The second phase includes investigation of its properties and relationships. The third phase includes the development of measurement units and methods. Although I've placed measurement in the third phase, I think it's necessary to make estimates in all phases, and I'm sure improvements will be made to the estimates when and as that becomes feasible.

Since my proposed research is in the first phase—establishing what exists—I will use estimates only to make choices in experiments. And I will not be concerned about accuracy or the properties of a thing except to design experiments to establish its existence. Since an astrologer, for example, is already confident that a given influence exists, he or she focuses on understanding the nature and intensity of that influence—what I call phase 2. Accuracy and properties are what phase 2 research is all about, but in phase 1, we only need to look for contrast. If two things are shown to be different from each other, contrary to the null hypothesis, this provides evidence that the null hypothesis is false. Having established that, we can move to phase 2. Thus, my suggested experiments frequently rely more on "contrast" than on quantitative or qualitative characteristics. This concept of contrast is important in the context of complex systems, and I will devote more time to its explanation later in this section.

Astrological influences are easier to study than other nonphysical phenomena because we can observe effects in the physical LOR. We have seen that the benefits of astrological research may significantly improve our lives, so I have placed some astrological experiments first. With properly designed experiments, we can establish some useful correlations without learning to observe the associated causal factors.

Some of the later experiments require direct observation of the causal factors and could lead to a better understanding of the influences on our lives, how we can change those influences, and also improve our understanding of the etheric and astral LORs.

Experiments in later chapters focus more directly on the etheric LOR.

Chapter 17: An Approach to Astrology Research

As nonphysical LOR topics go, astrology should probably be one of the later ones that we would investigate. In many ways, I suspect it is complex and advanced. I have two primary reasons for proposing astrology experiments at this early stage of the research.

First, it's relatively easy. No knowledge or awareness of nonphysical LORs is needed for initial experimentation because we can observe effects on the physical LOR.

Second, astrology is directly relevant to our lives precisely because the effects *can* be observed in our lives. [89] [90] As I explained earlier, I have been able to perceive some of the more extreme astrological influences on the etheric LOR using etheric vision. Based on this experience, I expect that scientists will eventually discover a causal relationship that involves etheric energy and has many effects on the physical LOR. But I'm getting way ahead of myself here.

My statements in this section are based on my experience. Some of these statements may seem extreme to you, and I do not ask you to "believe" them. I do, however, ask that before you discard them, you run my simple sun-sign experiment. If you use the formal version of the experiment, I am confident that you will see clear evidence as hypothesized. If you don't get

---

[89] Roderick, Main. (1998) *Jung on Synchronicity and the Paranormal* (p. 11) Princeton, NJ: Princeton University Press.
[90] Hippocrates (circa 460—circa 370 BC) multiple sources on the Internet; retrieved July 4, 2006.

definitive results, please publish your results and write me with any comments you may have concerning my suggested experiments and the results.

This chapter suggests some principles for initial scientific research in astrology. I feel that this is necessary due to the many beliefs and misunderstandings relating to astrology. Astrology has always been a part of our lives, not merely because its influences affect our lives, but because people have studied it for tens of thousands of years, even using it during the Inquisition. [91] Since astrology has not been studied as a field of mainstream science, no scientific standards have been established. As such, anyone can publish a book on astrology, and few people have sufficient knowledge of the field to say if it contains valid information.

This leads to my first suggested principle in doing astrology research—don't believe anything you read about astrology. I.e. question everything you read or hear. On the other hand, I have found that much of the astrology literature is valid to some extent. How can I say that? Personal observations, study and experimentation. This goes back to the basic scientific approach presented by Carl Sagan [92] regarding a balance between skepticism and openness, followed by experience gained by experimentation.

---

[91] Professor Teofilo F. Ruiz. Professor at the University of California at Los Angeles. Terror of History: Mystics, Heretics, and Witches in the Western Tradition course http://www.thegreatcourses.com/courses/terror-of-history-mystics-heretics-and-witches-in-the-western-tradition.html

[92] Sagan, Carl. (1996) *The Demon-Haunted World: Science as a Candle in the Dark* (p. 27). New York, NY: Random House.

On the basis of my experience with astrology, my second suggestion is that existing astrology literature be used as reference material to select and design scientific experiments. Allow me to present an example of this principle. The selection mechanism in my simple sun-sign experiment requires a subject to choose between three personality descriptions. These three descriptions are based on the person's sun-sign and those immediately before and after that sun-sign. Admittedly, I stumbled on this by accident, but now I understand why this experiment is so powerful, while similar experiments are inconclusive. The astrological signs are commonly categorized in three ways. First, a sign is considered to be male or female, and the nature of the sign reflects one or the other. The first sign (Aries) is "male," the second is "female," etc. Thus, the signs before and after a given sign are considered to be of the opposite sex. Second, each sign is considered to be "cardinal," "fixed" or "mutable." Aries is cardinal, the next sign is fixed, the next is mutable, and so on. Again, the nature of the sign reflects one of these three characteristics. Thus, the adjacent signs are of the other two types, and reflect those characteristics. The third attribution of the signs relates to the four "elements." Aries is fire, the next sign is earth, the next is air, the next is water, and so on.

If you compare two "water" signs, you will find some commonality between them. The same is true if you compare two "cardinal" signs or two "female" signs. The only consistent way that you can guarantee that none of these characteristics are shared between two signs is for them to be adjacent. This guarantees that where one sign is "male," the other is "female." Where one sign is "fixed," the other will either be "cardinal" or "mutable." Where one sign is "earth," the other sign will have one of the other elements attributed to it. My experiment compares the person's calculated sun-sign with those immediately before and after—precisely the two signs that have the least in common with that sun-sign. This is an example of what I am calling "contrast" in these experiments.

I know that "common wisdom" and mainstream science beliefs can make no sense of it, but these three types of attributions are common in western astrological systems, and are repeated in many books. Astrologers consider them to be clear and important. I'm not suggesting that you "believe" these attributions, but I am suggesting that commonly accepted astrological principles be used to design your experiments.

Perhaps I should give an analogy to physics. Suppose you were going to experiment with a magnetic field, testing the field characteristics when the current is varied in an electromagnet, but the electromagnet in question is part of a machine and it is not practical to separate it. You begin by running a known current through the electromagnet and you test the magnetic field strength. You increase the current and measure again. You repeat this several times. Unfortunately, you didn't notice that other magnets exist in the same machine. When you increased the current to one electromagnet, the current was decreased in another, and some are fixed magnets, generating constant magnetic fields. What have you learned? Not much. In fact, your results are probably confusing and contradictory. Each magnet influences the net magnetic field.

Naturally, a scientist would isolate the magnet and first test the initial magnetic field, which includes a component generated by the Earth. Unfortunately, at our current stage of understanding, we don't know how to isolate one astrological influence from all the others. Researchers in the field of psychology are used to this type of problem because you don't find people who have the one extreme psychological condition that you wish to study, and are otherwise normal. They deal with this limitation in two ways. First, they attempt to find people with as few complications as possible. Then they study a large number of people with the given condition. Statistically, they hope and expect the large number of people in a study to normalize all other factors so that the common condition is primarily responsible for the observed results.

The simple sun-sign experiment effectively isolates the effects of the sun-sign by using the adjacent signs for comparison—or does it? What are the other "magnets" lurking in the system? If you study astrology, you will find that the influences due to the positions of the planets and the moon can confuse the influence of the sun-sign. So can the "rising sign." In fact, conflicting influences can also come as a result of the angles between the planets, Sun and Moon (collectively called planets), known as "aspects." This is why I recommend that an experienced astrologer be asked to exclude some of the targets in the experiment. If this exclusion is done carefully, I expect that of the remaining targets, only a couple of percent will be attributed to an incorrect sign by subjects who know the targets well.

Some people have taken the opposite approach, trying to find the best match for a person among the twelve zodiac signs. Consider the analogy with the system of magnets. The influence of the sun on a person's personality is analogous to the influence of one of the magnets in the system. Suppose you select one electromagnet (corresponding to the sun in this analogy) and try to predict what current would be flowing through that electromagnet to create the observed magnetic field. Of course, whatever you predicted would often be wrong because there are other magnets operating independently in the system, contributing to the measured magnetic field. That is the situation in astrology. I.e. a personality is affected by many astrological influences, and the net result may appear more like a different sun-sign.

You can see that even in my simple experiment, it is not really as simple as it seems. Based on the claims of astrologers, the net astrological influence (part of which is reflected in the person's personality) is quite complex. My experiment utilizes contrast (i.e. differences between a sign and those adjoining it) to demonstrate the existence of consistent correlations. By comparing only the two adjacent signs with the target's calculated sun-sign, the subject is essentially presented with a red apple, a green apple and an orange, and is asked to judge which of the three choices is more similar to an orange. The astrology literature can be used as reference material to predict such contrasts, thus improving your experimental design and your results.

Astrology is usually thought of as a system of correlations. If the astrological factors are influences, as I claim them to be, then the correlations are the observed effects of those influences. However, no matter why the correlations occur, their existence is significant and observable. Thus, I suggest that your initial astrology hypotheses should make statements about the existence or absence of correlations. Again, this is a common theme in psychology, i.e. studying correlations. This is another reason to apply many of the psychology experimental design principles to astrology experiments.

You may wish to design experiments to test the use of astrology to predict events. I recommend against this point of view. If astrological correlations are in fact the effects of etheric level influences, as I have suggested, then those influences cannot *determine* future events. [93] On the other hand, people who are unaware of those influences will tend to react in relatively predictable ways, resulting in somewhat predictable events. Therefore, astrology only results in correlations between certain astrological patterns and certain types of events, when the affected people are ignorant of the astrological influences. That's my opinion, and I'm sticking with it ... until somebody comes up with a better explanation.

I suggest that we develop a foundational knowledge of astrology and build on that. By foundational, I mean the strongest and best known correlations. I.e. I'm suggesting that our early hypotheses be based on correlations that are stated to be strong in the astrology literature, with associated experiments that use contrast to make those correlations as clear and distinct as possible. Once we have a foundation of experiments that provide dramatic and consistent results, other subtler principles can be studied more easily.

If you use existing astrology literature for reference, as I suggest, I think that a corollary is that you could use a skilled astrologer for similar purposes. I highly recommend this, but at the same time, I would not *believe* what he or she said. A skilled astrologer might save you significant time and headaches in your experimental design.

---

[93] Hickey, Isabel M. (1992) *Astrology: A Cosmic Science: The Classic Work on Spiritual Astrology*. CRCS Publications.

Chapter 18: A Simple Sun-Sign Experiment

This experiment is probably one of the most interesting legs of our journey. If the results are as consistent as I expect, it represents a significant breakthrough, and will lead the way to discoveries that have not even been suspected by psychologists, other than C. G. Jung.

This chapter includes the following elements.

[1] The hypothesis

[2] An abbreviated version of the experiment

[3] A detailed experimental procedure designed to test the hypothesis

[4] An explanation of why this particular experimental procedure reduces bias and provides accurate evidence.

As you would expect, this experiment is designed to utilize contrast as much as possible. The specific qualities or intensities of the influences of the sun-signs are mostly ignored, since the purpose is to establish the existence of the sun-sign influences, or at least that a correlation exists between sun-signs and observable psychological characteristics. Later chapters suggest experiments that establish the causal factor behind the correlation shown here.

SUN SIGN CORRELATION HYPOTHESIS

A direct correlation exists between the sun-sign of a person and his (her) observable characteristics. That is, certain psychological characteristics are partially predictable based on the zodiac sign of the sun at the time of birth. This is not a claim as to the specific nature of those characteristics, only that such exist that are thus predictable from the sun-sign.

ABBREVIATED EXPERIMENT

I am including this abbreviated version of the experiment mostly for the purposes of clarification. This version does not include methods to minimize bias, and its results should not be construed as scientific evidence—merely suggestive. However, with its simplicity, I think it is indicative of how the structure of the full experiment is intended to work.

THE EXPERIMENT

Sit down with a one or more interested friends and/or spouse (partner). List all the relatives, friends and co-workers that you collectively know well, whose birthdays you also know. We will call these people the Targets.

Determine the sun-sign of each potential Target. If possible, calculate the degree in the sun-sign. (Each zodiac sign has 30 degrees, so the Target's sun will be at least 0 degrees and less than 30 degrees of the sun-sign.) If you know the year and day of birth, several web sites provide free astrological calculations. If you don't know the birth year, you can get a rough idea of the sun-sign using the birth date from the following table.

FROM	TO	SUN SIGN
3/24	4/18	Aries
4/24	5/18	Taurus
5/24	6/18	Gemini
6/24	7/18	Cancer
7/24	8/18	Leo

8/24	9/18	Virgo
9/24	10/18	Libra
10/24	11/18	Scorpio
11/24	12/18	Sagittarius
12/24	1/18	Capricorn
1/24	2/18	Aquarius
2/14	3/18	Pisces

Table 1: Sun Sign Estimation

Yes, roughly six days of each month are excluded. If you include them, your results may be weaker. Similarly, if you can calculate the degree of the Target's sun, only include those Targets whose sun is between 2 and 28 degrees of their sun-sign. The first and last degree of each sign is considered to be "on the cusp," and is considered to indicate a mixture of influence from the two adjacent zodiac signs.

From my table of personality types in Appendix A, read the descriptions of each Target's sun-sign, and the descriptions immediate before and after. Select the description that best fits the Target. A "hit" is when you selected the description of the actual sun-sign of the target. A "miss" is when you selected one of the other two descriptions. Be as objective and accurate as possible.

Tabulate all the hits and misses. If there were no correlation between sun-signs and personality types, you should have approximately ⅓ hits and ⅔ misses. You will probably find that nearly all of your selections were hits, but remember that this was not a scientific experiment. It was only indicative of the results you might get when you run the more controlled experiment.

Since you knew the "correct" description for each target person, your results were biased, no matter how hard you tried to be objective. If, for example, you had all hits, it might be due to your bias in favor of astrology. Similarly, if you had many misses, this might be indicative of a bias against astrology. Your bias must be removed from the experiment to determine the actual presence or absence of a correlation. A well-controlled experiment is necessary to achieve this, as described in the following experimental procedure.

EXPERIMENTAL PROCEDURE

This procedure includes five types of participants; namely, [1] Experimenter, [2] Assistants, [3] Targets, [4] Subjects and [5] Astrologer.

OVERVIEW

Following the procedure as set out below, the Experimenter selects the Subjects and Targets, separating the Targets into three groups—primary, excluded, and control.

For each selected Target, the Experimenter determines the date (including the year) of birth. The time and place of birth are useful, if available. The Experimenter may ask a skilled astrologer to use the birth information to identify those Targets who are likely to be misjudged, based on other astrological influences. Such Targets are placed in the excluded group.

For each Target, the experimenter selects three sun-sign-related personality descriptions and makes note of which description will be considered a "hit." An assistant gets these descriptions without any indication of which of the three descriptions is a "hit." The assistant provides the three descriptions to the Subject to read and select the one that most closely fits the Target's personality.

After all Subjects have judged all related Targets, the Experimenter collects and tabulates the "hits" for the Targets. Each group of Targets is tabulated and analyzed separately.

EXPERIMENTER ROLE

The Experimenter coordinates the application of the experimental procedure. You may wish to use a spreadsheet such as in Appendix F to keep track of related information.

The process of selecting and placing the Subjects and Targets may not be linear. You may choose to approach potential Subjects and select the Targets based on who the Subjects know. For example, you may select Subjects from among your friends and co-workers, and choose Targets from among their friends and relatives. Alternatively, you may approach potential Targets and select Subjects based on close associations with the Targets. For example, you might select Targets from among your co-workers or a class of students, and approach their relatives to be the Subjects. This latter approach has the advantage that you may be able to get their birth day and year from company or school records. Here are some primary considerations in making your choices.

The birth information for each Target should be as complete as feasible, without giving out too much information about the nature of the experiment. For example, if you can easily get people's birth day and year from school or company records, you may wish to make-do with that level of detail, thus choosing your Targets based on convenience. However, if you have access to an astrologer to help select the Targets for the excluded group, you may wish to determine the time and location of birth for as many Targets as possible.

The location of a Target's sun in the sun-sign is an important factor in selecting the group in which to place them—primary, excluded, or control. If at the time of birth, the sun was known to have been between 2 and 28 degrees of the Target's sun-sign, that Target can be used in any of the three groups, as deemed appropriate based on other considerations. However, if the sun may have been at less than 2 degrees or more than 28 degrees, I would not use the Target in the primary group. Furthermore, if the birth day or year is deemed untrustworthy, using the person in the experiment can reduce the accuracy of the results.

Ideally, each Subject knows each related Target well enough to accurately judge the Target's personality. This has two parts. One is how well the Subject knows the Target. The other is the ability of the Subject to judge someone's personality. You may not wish to use some people as Subjects if you feel they are incapable of judging personality traits.

In some scenarios, you may wish to use Subjects with unknown ability to judge personality types. In such situations, I suggest using three Subjects per Target, and use a voting scheme in the analysis. Specifically, I would consider a "hit" to be when at least 2 out of the 3 Subjects agree and select the "correct" personality description. I would consider all other choices to be a miss. This significantly reduces the probability of a hit according to the null-hypothesis, but if my claims are true, the results would be the same as if all the Subjects were good at judging personality types.

For each Target, the Experimenter selects three personality descriptions from Appendix A. For Targets in the primary and excluded groups, the descriptions are those for the Target's sun-sign and for the signs immediately before and after. The "correct" selection is the Target's known sun-sign. For Targets in the control group, the three descriptions are selected randomly from among the zodiac signs not matching the Target's sun-sign. One of those three descriptions is randomly chosen as the "correct" selection. Any regularity in the description selections for the control group may introduce bias in the results due to similarities to the sun-sign. However, you might decide that it is even more important to use consecutive descriptions so as not to provide any clues to the assistants or the Subjects. If so, I suggest using three consecutive signs where the center one is of the opposite polarity to the known sun-sign of the Target. This is simple, if you consider that the odd numbered signs are male and the even numbered signs are female, according to astrologers. E.g. if the known sun-sign is Taurus (number 2), then use an odd-numbered description, such as Libra (number 7), as the center of the three selected descriptions.

The three descriptions for a Target in the primary or excluded group are randomly shuffled to avoid predictability by the assistant or the Subject. Even with this shuffling of the descriptions, assistants are likely to recognize the pattern of the personality descriptions, especially for Targets in the primary and excluded groups. Thus, I suggest that you explain to your assistants that they should not give any hints to the Subjects.

Indirection when identifying the personality descriptions helps to avoid potential bias. I.e. if you included the numbers associated with the descriptions in Appendix A, the assistants would have enough information to bias the results. I recommend a separate table (e.g. in a spreadsheet such as in Appendix F) that identifies the three zodiac signs associated with the descriptions. The table would indicate the sign that is considered to be a hit. The personality descriptions are shuffled randomly and then labeled. This reduces the cues available to either the assistants or to the Subjects. The table would be known only to the Experimenter until all judging has been completed.

The Experimenter gives the three randomized descriptions to an assistant to present to the Subject.

The results for primary Targets are included in the statistical analysis to provide evidence for or against the hypothesis. The results for the other two groups of Targets are used to validate the experiment. The results from the control group are expected to be consistent with the null-hypothesis, showing no correlation. The results from the excluded group are expected to contradict the null-hypothesis, but not as strongly as those of the primary group.

The analysis of each group of Targets is based on the “null-hypothesis” that there is no correlation. I.e. the Experimenter calculates the probability that the results occurred by chance, assuming that no correlation exists. My prediction is that the probability of the results occurring by chance will be—very small for the primary Targets, slightly higher for the excluded Targets, and quite high for the control Targets.

SELECTION OF SUBJECTS

The Subjects perform the key operation of judging the personality types of the Targets.

NOTE: If the Subjects are allowed to know that astrology is involved before the experiment has been completed, it could be construed as biasing the outcome of the experiment.

The Experimenter may select multiple Subjects for each Target. If three Subjects can be found per Target, a voting scheme can be used among the Subjects for each Target, as previously mentioned. However, a chi-square analysis can be used in a relatively straightforward manner with any number of Subjects per Target. Each Subject-Target pair can be thought of as a single and independent trial. Accordingly, I recommend that the Subjects associated to a given Target do their judging independently so that they do not influence each others' opinions. This requires a separate session for each Subject.

Preferably, the Subjects have minimal knowledge of astrology, and they are led to think that the experiment is about personality types. One approach to accomplish this may be to give them a survey, such as provided in Appendix B, Part 1. The point of the survey is to be misleading, while determining the person's degree of knowledge of astrology. Notice that the first few questions are, in fact, relevant to the experiment, while the question about astrology is included among topics that are currently ridiculed in western society.

If they need more information, you might tell the Subjects that you are attempting to learn more about personality types and how they are perceived.

If you find strong evidence of the hypothesized correlation, this statement is actually true. I.e. it will be evidence about the perception of the personality types associated with the sun-signs.

I think the Subjects should have at least average intelligence, and I would not include any who show signs of poor judgment when dealing with other people.

Naturally, I can't tell you how to select Subjects in your specific situation, but I do have some ideas. For example, if you are teaching a college class, you might ask your students to participate as Targets, and ask their parents and siblings to participate as Subjects. The student's birth date could be obtained from the school records. The surveys and personality descriptions could be sent by mail to the parents and siblings of each student, eliminating the need for assistants as intermediaries. Using voting with three Subjects per Target would be useful in this scenario because you have no information about the quality of judgment of the Subjects.

Alternatively, if you wish to run the experiment at a business, you might enlist co-workers as Targets and/or Subjects. Spouses, children and parents might be included as Targets and/or Subjects. If you are a manager, you probably have records of the birth data for each employee, making them prime targets, ah good choices as Targets. Again, if you do not personally know the Subjects, a voting scheme is advisable.

If you are short on Subjects, you might wish to include astrology-knowledgeable Subjects for the control Target group.

TARGET SELECTION

These are the people whose personality types will be judged by the Subjects. I recommend enough Targets to provide at least 60 data points, i.e. Subject-Target combinations. More data points are better than fewer. As previously discussed, birth data (day, year, and preferably time and place) is needed to choose the group for the Target. If only the birthday is known, the Target can be included in the control group, as long as care is taken to select personality descriptions that cannot derive from his or her sun-sign.

NOTE: Many people will give a false age or birth day on a survey. This could bias your research, so some form of verification is advisable.

The birth information of the Targets in the control group must be sufficiently well-known to be sure that none of the choices are valid, and I suggest a control group with at least 20 Targets.

If you do not have a convenient way to get birth information, you may wish to give a survey such as in Appendix C, Target Survey. The primary purpose of this survey is to obtain the Target's date of birth, but it also misleads him or her into thinking that the date of birth is not critical information, and by inference, that you are not performing an experiment related to astrology.

THE ASTROLOGER'S PART

An astrologer can use the birth information of Targets to exclude those deemed to have influences that would confuse the judging Subjects. Specifically, ask the astrologer to identify all subjects where there might be confusion about the personality type between the birth sun-sign and the two adjacent signs. E.g. if the person is an Aries, the astrologer should check if the person might be confused as having a Pisces or a Taurus sun-sign. To do this, the astrologer will check for strong conflicting influences from the positions of the other planets or their aspects to each other. A good astrologer will understand. If he or she doesn't understand the request, find another astrologer.

The Targets identified by the astrologer are placed in the excluded group.

ASSISTANTS

Assistants handle the interactions with the Subjects during the judging phase of the experiment. Since the assistants do not know the "correct" personality description for each Target, the procedure is called "double-blind." In this part of the experiment, assistants interview each Subject, and record his or her decision for each Target he or she has been selected to judge.

You may wish to record these interviews with both audio and video. The recording may be useful to reduce criticism from those who would accuse you of introducing bias into the experiment.

A script may be advisable for the interactions with the Subjects. For example, the assistant might say something like, "I'd like you to read these three personality descriptions and select the one that most closely matches this person's personality." The primary point is to introduce as little bias as possible. Scripted responses may also be advisable for certain expected questions from the Subjects, such as, "Why am I doing this?"

ANALYSIS

I recommend the use of a chi-square analysis to estimate the probability that the experimental results happened by chance. At the time of writing this, you can find at least one web site [94] that will calculate the chi-square value for you. Chi-square values greater than eight (8) are indicative that a correlation does exist, but the associated probability would also be of interest in your published results.

EXPLANATION / JUSTIFICATION

Much is claimed by astrologers concerning correlations between astronomical configurations at the time of birth and influences on an individual. By using the common attributions given by astrologers, we can probably improve the effectiveness of our experiments. For example, each sun-sign has little in common with either the sun-sign immediately before it in the Zodiac or immediately after it. I explained this in some detail earlier.

Furthermore, astrologers claim that if the calculated position of the sun is close to the "cusp" of the sign (zero or 30 degrees of the sign), the expected characteristics of the individual may be confused with those of the sign on the other side of the cusp. Thus, a Target should only be included in the primary group of a study if the sun was between 2 and 28 degrees of their sun-sign at the time of birth.

The experimental procedure includes a control group. This is common in scientific experiments. The primary purpose of the control Targets is to

---

[94] http://home.ubalt.edu/ntsbarsh/Business-stat/otherapplets/goodness.htm

demonstrate that the selection of the "correct" personality type by the Experimenter did not bias the judging by the Subjects. More generally, if the same results were obtained in the control group as in the primary group, it would be an indication that the experiment did not accurately test the hypothesis, or that the hypothesis is false.

I included the "excluded" group in this procedure primarily because astrology is so controversial, and related experiments are more likely to elicit criticism. One of the criticisms may be that a Subject's conscious or subconscious knowledge of the Target's sun-sign may influence the Subject to select the "correct" personality description. If this criticism were a significant factor in the results, then the excluded group would have the same rate of hits as the primary group. Although the rate might be equal between the two groups in a small study, in larger group sizes such as over 100 Targets per group (with proper exclusion), I expect that a higher hit rate will be found in the primary group, thus providing evidence that the criticism is false.

The role of the astrologer is to apply astrology-specific knowledge to reduce the chances of incorrect selections by the Subjects. If the astrologer excludes too many Targets, the excluded group will have nearly the same hit rate as the primary group. That might increase the statistical significance of the results for the primary group, but it would reduce the usefulness of the excluded group.

Chapter 19: Experiments with Astrological Transits

This chapter suggests a family of experiments to develop evidence of astrological influences from events called "transits." A transit represents a type of astrological influence from the planets (i.e. the Sun, Moon and planets) at the current time. In contrast, the last chapter describes an experiment related to the position of the Sun at the time of a person's birth. A person's astrological chart at birth is known as the "natal" chart. The natal chart shows the positions of the Sun, Moon, Mercury, Venus, Mars, Jupiter, Saturn, Neptune, and Pluto (yes, I know Pluto is now considered to be a dwarf planet). A natal chart typically shows many relationships and principles, depending on the background and training of the person creating the chart. Besides the positions of the planets at the time of birth, a typical natal chart includes "aspects." An aspect is a meaningful angle between planets in the natal chart. Most angles are considered to be unimportant, but when the angle between planets is close to 0° (zero degrees), 60°, 90°, 120°, or 180°, astrologers expect significant influences.

While an aspect is a meaningful angle between planets at the time of birth, a transit is a meaningful angle between a planet in its position at the current time (the transiting planet), and the location of a planet at the time of birth (the transited planet). For example, a person is having a transit if Mars was at 21° 17" (21 degrees 17 minutes) of Scorpio at the time of birth, and Uranus is currently at 21° 15" of Aquarius. This is an angle of 89° 58" (i.e. close to 90°). A transit is normally considered to be in effect only when the current angle is within one degree, as shown in the following table.

NAME	EXACT ANGLE	MINIMUM	MAXIMUM
Conjunction	0°	359°	1°
Sextile	60°	59°	61°
Square	90°	89°	91°
Trine	120°	119°	121°
Opposition	180°	179°	181°

Table 2: Transit Angles

Astrologers are often interested in other angles. I have selected these specific angles because they indicate the strongest influences, and are therefore the easiest to observe in the context of an experiment. A 90° angle is called a square. An angle of 180° is called an opposition. Angles of 60° and 120° are called sextiles and trines, respectively. 0° is called a conjunction. In general, squares and oppositions are influences that tend to increase difficulty while sextiles and trines make life easier. The effect of a conjunction depends on the context—some make life easier while others make life more difficult.

I will not attempt to describe all the major types of transits. I suggest additional reading for those who are interested in designing other experiments to study the effects of transits. I recommend *Astrology, a Cosmic Science: The Classic Work on Spiritual Astrology* by Isabel Hickey [95] as a primary reference to learn more about most influences, from an astrologer's perspective. For additional information about transits, I recommend *Planets in Transit: Life Cycles for Living* by Robert Hand. [96] Again, I suggest the use of these books only for reference, but I contend that to ignore such writings courts poor experimental designs.

---

[95] Hickey, Isabel M. (1992) *Astrology: A Cosmic Science: The Classic Work on Spiritual Astrology*. CRCS Publications.

[96] Hand, Robert. (2002) *Planets in Transit: Life Cycles for Living*. Whitford Press, U.S.

When dealing with astrological influences, always keep in mind that each individual may be affected by dozens of such influences, most of which are too minor to notice. This complex pattern of influences creates the need for contrast when attempting to establish the existence of a particular influence. But it also points out the need to quantify the intensity of each astrological influence. A rough estimate of intensity is needed in order to select which transits to consider and which can be ignored. This process of estimation is quite complex, and if you can enlist the aid of a professional astrologer to help make the estimates, I highly recommend it. In lieu of professional assistance, I'll try to give some guidelines.

We know how to quantify the intensity of a magnetic field—the most prevalent unit is a "Telsa," whereas an older unit for measuring magnetic field intensity was a "gauss." We can measure water currents in terms of gallons per minute, with the intensity often measured in pounds per square inch. But what units make sense for astrological influences? Some influences are almost irresistible, while others are so weak that no one would be affected.

How do we define the units? How do we take measurements? The following is an inaccurate method of measurement for astrological influences. I simply suggest it as a starting point, until a better approach is discovered.

For this purpose, I suggest the unit of intensity for measuring astrological influence be that of the Sun when it is at 5° of Leo. The intensity of this influence is not as strong as some, but it is easily observed in the context of suitable contrast, as can be seen from the results of the simple sun-sign experiment.

Presently our best basis for comparison of astrological influences may be books such as Isabel Hickey's. It is my hope that research will uncover better units and methods for measurement. I think some of the experiments I suggest in later chapters may lead to such knowledge, but for now, please bear with me.

The influence of the Sun is strong in Leo, but it is strongest in Aries. The Sun's influence is weaker when in Aquarius, and weakest in Libra. For this reason, the Sun is said to "rule" Leo, be "exalted" in Aries, in its "detriment" in Aquarius, and in its "fall" in Libra. To learn more about these terms and their application, I recommend Isabel Hickey's *Astrology: A Cosmic Science: The Classic Work on Spiritual Astrology*, as mentioned earlier. For lack of a more accurate comparison, I suggest the following table of relative intensities as a starting point. I hope that you will contribute and help improve these numbers.

SIGN	INTENSITY
Aries	1.2
Taurus	0.8
Gemini	0.9
Cancer	0.8
Leo	1
Virgo	0.8
Libra	0.5
Scorpio	0.9
Sagittarius	0.9
Capricorn	0.9
Aquarius	0.7
Pisces	0.8

Table 3: Sun Sign Intensities

Using these factors relative to the intensity of the influence of the Sun at 5° of Leo, we have an initial (extremely rough) estimate of the intensity of the Sun in each sign. For simplicity, I'll call the unit of measure a "Leo-Sun." I will use this unit for the purposes of comparing transits in this chapter.

Remember that a Leo-Sun is a unit of intensity. We will see later in this chapter that the influence of a transit can be cumulative. When it is, a 0.5 Leo-Sun transit can manifest as if it had an intensity of 1.5 Leo-Sun. For our present purposes, I will ignore this inconsistency, and estimate the intensity of a transit based on the intensity of its typical manifestations. Hopefully, later research will straighten this out.

As mentioned earlier, some transits usually have positive effects while others typically have negative effects. For this discussion, let's say that minor transits have intensity less than 0.2 Leo-Sun and major transits have intensity at least 1.0 Leo-Sun. Only major transits will be recommended for the current experiments. Transits with intensities greater than 0.2 and less than 1.0 are not strong enough to provide clear contrast, but can influence events, potentially confusing the results of our experiments.

Here are some guidelines and examples on how to estimate the intensity of transits. First, if the transiting planet is one of the fastest, consider the transit to be minor. These "planets" are the Sun, Moon, Mercury and Venus. Mars and Jupiter transits may be significant, but are not consistently major. If the predicted influences of these transits are consistent with a targeted major transit, no problem. But if some are negative while others are positive, I suggest discarding measurements taken when Mars and Jupiter transits are active, or within a week after the end of one of these transits. Transits by Saturn, Uranus, Neptune and Pluto are usually significant and are often major. The intensity of a transit depends on the transiting planet, the angle of the transit, and the strength of the transited planet.

Part of the strength of a natal planet relates to the sign it is in. If it is in the sign it "rules" or in which it is "exalted," it is strong. If the planet is in its "fall" or "detriment" sign, it is weak. Otherwise, it can be considered to be of average strength. Another part of its strength relates to its aspects (meaningful angles) to other planets in the natal chart. If it has strong aspects to strongly placed planets, such aspects increase its strength. If a planet is in its detriment (or "fall") sign and has no strong aspects to strongly placed planets, the planet is probably weak.

Transits with an angle of 30° (known as a semi-sextile) can be considered to be minor. If the transiting planet is Saturn, Uranus, Neptune or Pluto, and the angle is 45° or 150°, it may be significant, and should be avoided when the predicted effect contradicts the effect of a target major transit.

A major transit then is one satisfying the following rules:

[1] the transiting planet is Saturn, Uranus, Neptune or Pluto,

[2] the transited planet is not weak in the natal chart, and

[3] the transit angle is 60°, 90°, 120°, or 180°.

Now that we've established some general rules to estimate the intensity of a transit, we'll complicate it a bit more. You will find that transits sometimes come in groups. If there is a close aspect between two natal planets, then when one is transited, so is the other. For example, if Mars forms a 59° sextile with the Moon (both in the natal chart), then when Uranus transits trine Mars (recall that a trine is 120°), Uranus also transits either sextile (120°-60° = 60°) or in opposition (120°+60° = 180°) to the Moon. In the second case, a square is negative while a trine is positive, and the net effect may be difficult to predict. Thus, in the presence of contradictory major transits, I recommend discarding the data.

These rules are generalizations, and although usually good enough, they may not always hold true. Again, your best bet is to consult with a professional astrologer to estimate the level of significance of the transits you might consider including in a study. I also recommend that you read the previously cited astrological references by Isabel Hickey and Robert Hand for a better understanding of transits.

Although I often reference other books on the topic, as you have seen, I find it necessary to explain some characteristics of astrology as a basis for my reasoning, and for the proposed experimental designs. As I've said, please do not *believe* what you read (i.e. maintain reasonable skepticism), but it seems only reasonable to design experiments so that *if* these things are true, your experiments will be successful.

Intensity is one factor in determining the amount of effect a transit has on a person's life. Another factor is the duration. On average, the influence of a transit seems to be analogous to an unexpected influx of water in a home. Let's say a pipe leaks. If water pours in for a few seconds and stops, the effect is likely to be minimal. On the other hand, if you were on vacation and a minor leak allowed a drizzle into your living room for several weeks, the damage could be major. By the same token, if that drizzle went into your basement near a drain, there might be no damage at all. I.e. if the water builds up, the effect increases.

For the sake of discussion, let's say that the influence of a transit deposits energy (which I have observed in people's auras). A portion of that energy is "used" to produce an effect of the transit. If the energy is blocked, it accumulates until it finds a channel. Thus, if the energy is released slowly, the effects are hardly observable. If, however, the energy accumulates until a massive release occurs, the resulting effects are usually also massive and easily observable.

If a transit lasts only a few hours, the potential energy build up is minimal, and its effects will be difficult to evaluate (like a water pipe leaking for a few seconds). As a rule of thumb, I suggest targeting transits that involve only the slowest moving planets—Saturn, Uranus, Neptune and Pluto. Many of those transits last long enough and are intense enough to provide lots of energy to produce observable effects.

So at this point, we have the outer planets moving through the zodiac signs, occasionally transiting natal planets. If the intensity is significant for a long enough time, the energy is likely to build up and produce observable effects. The outer planets move slowly enough that the duration of a transit is usually enough to serve our present purposes. The descriptions of transits as found in Robert Hand's *Planets in Transit: Life Cycles for Living* are based largely on the current normal state of ignorance of nonphysical LORs in general and of astrology in particular. If someone is aware of such influences or able to directly perceive them, they may choose to channel the energy in an unusual manner, resulting in different effects. Such people are probably not good research subjects at this stage of investigation, and as more people learn about these influences and begin to take more control of the effects, astrology experiments will need to change accordingly. But for now, I simply mention it as a cautionary note.

As before, I suggest the use of contrast to highlight the presence of influences. At least two types of contrast are readily available when observing the effects of transits. One is the contrast between the presence of a major transit and the absence of any major transits. The other, and possibly more useful contrast, is between a major positive transit and a major negative transit. Let's look at a potential scenario.

A convenient venue for this type of study is a work place or college course. Let's use a college course of about twenty students as the basis for a sample experiment. One nice thing about most college classes is that the students are mostly of about the same age. This fact often implies that several students will go through some of the same transits during a three or four month period, such as a college term or semester. Also during the same period, many of the students will typically go through several different major transits.

The planets closest to the Sun move more rapidly than those farther out. As such, the slowest moving planets (as before, Saturn, Uranus, Neptune and Pluto) will have about the same positions in the students' natal charts. Thus, when a transit to a slow-moving planet affects one of those students, it usually affects others during a single semester. This presents the opportunity to study the effects of such a transit. If some of the students are not concurrently affected by the transit, the effects can be studied in contrast to those who are unaffected at the time.

Since the Sun, Moon, and other relatively fast-moving planets change their positions rapidly, a major transit to the Moon, for example, will probably only affect one of the twenty students during a semester. With only a small number of the students being affected by a transit at the same time, the situation offers contrast not only between those having a transit and those not having the transit, but also between positive and negative transits. This presence of positive transits for one student while another has negative transits provides the highest contrast.

Let's assume that this is an undergraduate psychology class. The professor could proceed as follows:

[1] Obtain the birth information for each student. (If the time of birth is missing, the location of the moon cannot be known accurately enough to use it as a transited planet.)

[2] Calculate the natal chart of each student.

[3] Calculate the transits that will occur during the course.

[4] Choose which contrasting transits to include.

[5] Devise simple questionnaires that will highlight the effects of the contrasting transits, probably on a pair-wise basis, and producing numeric results.

[6] Devise a questionnaire that will tell you how much each student knows about astrology, but without indicating the nature of your study.

The numeric results of the questionnaires will be easiest to use if they normalize nicely for each transit. I.e. if the net effect measured for each transit has a linear value ranging between 0 and 10 (or some other fixed range), the values can be compared more easily between transits. If the effects of a transit would result in negative values (e.g. the effects of difficult/negative transits), keep this in mind and invert the sign later, when the results are combined.

Having done this preparatory work, give the initial questionnaire to determine their knowledge of astrology. I suggest hiding the topic of astrology with mild ridicule. E.g. ask questions about their level of knowledge about urban legends, the Loch Ness Monster, ghosts, yetis and astrology. E.g. for each topic, ask them to specify their level of knowledge between 0 and 10. Since you may want to run similar experiments in multiple semesters, if you tell the students that you were investigating anything about astrology, it would tip off other students.

You might give the main questionnaires once per week. A cover story of some sort could reduce the likelihood that students will look at their own transits and respond to the influences in a more considered manner.

The questionnaires could include questions that the students would be likely to answer honestly, or if not honestly, then at least indicatively. If a transit is expected to bring violence into the student's life, some of the questions would illuminate the level of violence in the student's thoughts, feelings, reactions, and especially surrounding the student. For example, "Have you had any significant accidents in the last week? If so, please rate them on a scale from 0 to 10, where 10 is the most extreme." If a transit is expected to influence the student to be depressed, related questions could bring out the level of depression versus happiness.

For these purposes, I think a single questionnaire would be given to two students who will be used for the purposes of contrast, and also to students from a control group (students currently having no significant transits, either positive or negative). E.g. if one student is going to have Neptune transiting trine Venus while another student will have Uranus transiting square Mars, the first transit typically results in more romance and/or an increase in creativity, especially in some form of art, while the second transit typically results in confrontations and accidents. As such, the questionnaire could include questions about the students' love lives, their artistic inclinations, confrontations they experience, and accidents they have.

In each case, the questions would target not just the general principle, but the specific nature of the expected influence. For example, if a transit is expected to evoke violent reactions from people around the student, then questions would delineate the degree and frequency of such reactions. I.e. general questions about violence would not be as discerning, and would therefore not provide as much contrast. If a transit is expected to bring depression due to self-doubt, then questions would specifically dig into the level of self-confidence and the degree of happiness.

I would like to give a general caution here. Surveys and questionnaires already exist to study personality traits, depression, anger, etc. I suggest that these existing tools *may* measure principles that are orthogonal to the influences of the student transits. That is why I am suggesting the development of questions that specifically target the influences of the transits as predicted by astrologers. For example, anger is often linked to violence.

But if a transit does not involve anger, then including questions to determine the level of anger will be misleading at best, and would certainly reduce the contrast the experiment is attempting to show. Suppose, for example, that a transit is predicted to bring violence in the form of accidents and anger directed at the student. Some people react to anger with anger, while others cower away. Thus, the degree of anger in the student is mostly irrelevant (i.e. orthogonal) to the expected influence of the transit.

Basically, I am suggesting that the questions used in the experiment be carefully tailored to the specific influences rather than using well-established tools. And if the questions you select do not highlight the influence of the transits, there are at least three possible reasons.

[1] The questions were ineffective in showing the contrast,

[2] the influence didn't exist, or

[3] the students reacted to the transit influences atypically.

For major transits, I contend that the influences always exist. That's why I recommend these experiments—I'm confident that you can get significant results from a well-designed and well-run experiment. Although it is always possible for the students to react to the transit influences in atypical ways, in the absence of knowledge, that is unlikely.

So where are we at now? We have a class of students, some of whom are having powerful transits this semester during your psychology course. These may be transits involving faster transited planets (the Sun, Moon, Mercury, Venus) in the students' natal charts and slower transiting planets (Saturn,

Uranus, Neptune, Pluto) in their current positions during your psychology course. Thus, you've selected a subset of the students who are having major negative transits without significant positive transits, a subset who are having major positive transits without significant negative transits, and a third subset (the control group) who are not having any significant transits during some portions of the semester. You've devised questionnaires to give to the students on a weekly basis, probably pairing questions for one of the positive transits with questions for one of the negative transits on each questionnaire. On a weekly basis, you've given each questionnaire to the target students. This gives you a sequence of weekly numeric values for each target student and for each of the two transits for which each questionnaire was designed.

Now, what do we do with the data? The idea is to use contrast to provide evidence of the existence of transit influences. Two types of contrast exist in the data. First, there is contrast between the period before a transit begins and the period of the transit. If my statement is true that energy builds up during a transit, then we can't use the data after the transit ends. We could not be sure if residual transit energy might cause phenomena after the end of a transit. The second type of contrast is between the positive and negative transits.

Although curve fitting analysis may seem logical here, I suggest that it does not match the characteristics of the problem because the effects of many transits may be seen at any time during or after the period of the transit. While the effects of some transits will be obvious near the beginning and throughout the transit (suggesting a curve-fitting analysis method), the effects of other transits may only be seen in a single dramatic incident. Thus, I suggest analysis that is similar to measuring the area under a curve. My suggestion is as follows.

For each target student, add the results for each transit prior to the onset of the transit. Divide by the number of weeks. This is the base level for that student. Add the corresponding results during the transit. Divide by the number of weeks. This is the affected level for the student. Subtract the base level from the affected level. This is the observed normalized net effect of the transit. (If the predicted effect of the transit would result in a negative difference, invert the sign.) Having done this for each student, combine the results for the two students having the paired positive and negative transits, by adding their observed net effects. This is an application of contrast. Let's call this value O, meaning a single observation. Use the same approach for the control group students. For this purpose, I suggest treating a control student's data as if he or she had had the same transit as one of those for which the associated questionnaire was designed.

The null-hypothesis is that no transit influences exist. If the null-hypothesis is true, then the observed net effect of all transits will be roughly zero, the same as for the control group.

Consider each sample (O_1, O_2, . . .) to be the value of a random variable. The null-hypothesis implies that the average for each is zero. Assuming that the weekly average values range between minus 5 and plus five, the variance is less than 12.5 (i.e. 25/2). Now add all the observations together $T = O_1 + O_2 + \ldots$

If we add the values of N random variables, each of whose average is zero and variance is less than 12.5, the average of the sum is zero, and the variance is less than 12.5 divided by N. This is consistent with the weak law of large numbers, and provides a basis for analysis. If the null-hypothesis is true, adding more net effects together will reduce the variance linearly. In fact, I expect that to happen with the control group. If, however, my hypothesis is true that transits do influence and have observable effects, then T will grow as more observations are added.

We could improve the accuracy of this analysis by estimating the variance of the measurements, e.g. looking at the weekly scores. This is simple statistics, and I'll skip the math here. Let's assume you calculated the variances of the measurements to be σ_1^2, σ_2^2, σ_3^2, . . . A conservative estimate of the variance would be the maximum of these values. Let's call this estimated variance σ_e^2. σ_e^2 is likely to be much less than 12.5, allowing us to use a smaller sample size to produce significant results.

The law of large numbers also states that the statistical distribution of the sum of a large enough collection of random variables, all having the same mean and variance, can be estimated by the Normal Distribution. "Large enough" probably means more than 100 pairs of transits. Hopefully, we don't need that many samples to be convincing. I suggest that a value of T greater than $4\sigma_e$ provides evidence that the null hypothesis is false.

By the nature of this experiment, you can run it on as many students as you wish, either in parallel (courses given in the same semester) or series (successive semesters). I predict that a small number, e.g. 10, pairs of major transits would provide evidence of the existence of transit influences.

Chapter 20: Astrology: Phase 1 Complete?

As it has commonly been studied and practiced, astrology is a set of observed and/or predicted correlations between astronomical phenomena and physical/psychological phenomena. Discarded by mainstream science, astrology has been practiced consistently by people at all levels of society, and, yes, among scientists as well, because it works. The simple sun-sign experiment (hopefully) provides clear evidence that at least some of the correlations of astrology do work. Although the transit experiments are much more complex, I expect them, too, to provide such evidence.

If you are an astrologer, you probably didn't run the experiments because you're already confident of the outcome. If you are a scientist, I hope that you ran, or will run, at least one of these experiments, and publish the results, including the details of the experimental procedure you used.

As I am confident of the results of these experiments, I will assume that now you and I agree that some astrological correlations are valid. We might say that phase 1 of our research has only just begun—gathering evidence that these correlations work—since so very many correlations have been claimed true. But I suggest an entirely different approach.

Now that we agree that at least a few correlations exist, I claim that the next step is to find out what causes them and why they function as they do. Once we understand the causal relationships, we can use that knowledge to predict and test other correlations.

So where do we look for this causal factor? Sorry for the lack of suspense here. I already answered that question. I have observed the buildup of energy in people's auras during major transits. When that energy has been in my own aura due to a current transit, I have sometimes manipulated the energy to achieve a different outcome. As described earlier, I have also assisted others to do the same. I am confident that this energy buildup in the aura is part of the causal relationship that we can observe physically as a set of correlations.

On this basis, I suggest that we investigate such energy associated with transits, and in a later chapter I describe an experiment as an initial approach to that investigation.

Next step—find out what exists on the etheric LOR.

Chapter 21: Experiments with Auras

The basic astrological influences are easy to study because they affect our lives in obvious ways. When you learn what to look for, you can see the effects everywhere. But how do they work? What do they actually influence? *I* said they act through our auras using a type of energy, but we have not yet established that auras exist. This chapter describes an experiment to provide evidence that humans do in fact have auras.

Auras are easy to find and maybe the easiest etheric-LOR objects to observe directly. As such, they are an obvious starting point for fundamental research. However, it must be kept in mind that an aura may be complex, so the results obtained must be interpreted in that light. Experimenting on an aura may be analogous to experimenting on an animal to observe basic physical principles. In fact, the study of basic physical principles can best be made on inanimate objects such as a known mass of pure iron. Studying the inertia, momentum, and reaction to applied force is more complicated when experimenting with an animal, due to the animal's reactions and pliability. We do not yet have enough knowledge of the etheric LOR to know which objects might be equivalent to an animal and which might be equivalent to a bar of iron. It is only with the nonlinear development of scientific understanding that we are able to know the factors that influence our experiments and in what ways, so when I suggest starting with auras, I am not suggesting that it is optimal. I am only suggesting that it is convenient.

We have several ways to proceed in this research. As I have already given arguments on this topic, I will simply reiterate my opinion—the researcher should learn to make direct observations as soon as possible, even if one or more assistants are already skilled at etheric vision. In fact, I recommend three observers for each experiment. One reason for this is obvious—if two of them agree on their observations, it lends credence. I've mentioned the other reason elsewhere in this book—when two or more people work together in observing the etheric or astral LOR, their individual clarity and accuracy improves. If you are the only person capable of making observations with your etheric vision, then so be it. The experiments will work the same, but the results will be less compelling.

As with physical LOR experiments, mechanical enhancements, such as cameras, are always a good idea. Kirlian photography is reputed to make a physical record of an aura. Assuming that direct observations can be made with etheric vision, this Kirlian claim should be relatively easy to test. The test is like testing to see if a normal camera is functional. Look (in this case using etheric vision) at the object (in this case an aura) and take a picture. Draw a picture of what you saw, and compare the drawing with the photograph. If the drawing is a reasonable facsimile of the photograph, then the camera (in this case, the Kirlian equipment) is functional.

I assume that this test has been performed many times. Unfortunately, probably not by those who were the end-result of a 500-year indoctrination. My hope is that scientists who read the first part of this book now realize how they were misled, and will be interested in correcting the situation for themselves, applying the scientific method to investigate the forbidden topics. Hopefully, you are developing etheric vision and will run this test for yourself. I have not tested Kirlian photography, so I only know what I've read. However, I've seen some pretty convincing evidence, and that's why I recommend considering it. If Kirlian photography works, it could help immensely in the following experiment.

NOTE: This experiment, and most of the following experiments involve observers, some of whom may be using the AWIN system to achieve etheric vision. Always keep in mind the part of the training related to the vibration rate. While making an observation, always shift your vibration rate up and down until the intended target (e.g. the aura) is observed with maximal clarity. If the most effective frequency is unexpectedly high, make note of the effective frequency, as it may indicate that the object being viewed is on a higher LOR than expected.

HYPOTHESIS

A human has something that envelopes him or her on the etheric LOR, commonly called an "aura."

EXPERIMENTAL PROCEDURE

A room is selected where the participants will not be disturbed, and where the level of sound can be controlled (kept quiet).

A suitable subject stands in the middle of the room.

Kirlian photographic equipment is set up to record an image of the subject's aura when the subject chooses to do so. Preferably, at least three separate images can be taken per experiment.

One to three observers have comfortable positions along one wall of the room with nothing separating them from the subject. If necessary, observers may be located in other rooms or buildings.

The observers prepare to observe the subject's aura.

The subject changes his or her aura in a distinctive way—shape, size and/or position.

When the change is in effect, the subject notifies the observers to remember what they observe, and a Kirlian photograph is taken.

When finished, the subject notifies the observers.

Each observer makes notes of what he or she observed, including drawings and other descriptions to identify the size, shape and location of the subject's aura.

The subject makes corresponding notes of his/her intended aura changes.

The observers and the subject compare notes.

If Kirlian photographs are taken, compare these with the observations and with the intended aura changes of the subject.

METHOD 1

A subject is selected who has had AWIN training, including Meditation Method 3 and the Comet Exercise. With this training, the subject should be able to modify his or her aura.

Again, I suggest three observers with AWIN training or RV training. However, one observer is sufficient. If using RV training, verbalization may be necessary, requiring separation of the observers, probably in different buildings. If so, PCs may be used with instant messaging to coordinate the observations.

Kirlian photography may be omitted if it is considered ineffective or is unavailable.

The room contains no plants or animals, as their auras might be confusing to the observers.

The sound level is adjusted to suit the needs of the observers to promote concentration and accurate etheric vision.

No hints or unnecessary discussion is present that might bias the observers.

If Kirlian photographs are taken, they may be used to fine-tune the experiment. For example, if the photographs differ from the intended changes of the subject, choose a different subject. If some of the observers disagree with the Kirlian photographs, either switch observers or help them to use better techniques.

If a numeric probability is desired as an outcome from the experiment, a collection of aura changes may be pre-selected. In this case, the subject chooses one of this set of changes. For example, the height of the aura might be half, the same or double the height of the subject. The shape might be spherical, flattish or tall and thin. The location might be a few feet to the right or left of the subject's physical body. This would give 18 options for the subject to attempt, and for the observers to select from, based on their observations. The probability of all three observers randomly selecting the same option as selected by the subject is $1/18^3$, meaning 1 out of 5832. If three different successive options in an experiment are selected with the same result, the probability would be about $5.04X10^{-12}$. In other words, it would be dramatic evidence. If only two of the three observers agreed on the "correct" pattern, the probability would be about $2.48X10^{-8}$. So even if one of the observers disagreed on every observation, the results could still be highly significant. This is an example of how to get a numerical result from the experiment.

METHOD 2

This is the same as Method 1, except that a cloth screen is used to separate the observers from the subject, and a doorway is convenient, allowing the subject to leave the room without giving any physical queues to the observers. The point is that the observers make their observations and the Kirlian photos are taken whether the subject is present or not. The subject randomly selects to stay or leave. In the scenario described above, this provides a 19th option for the subject, which can be included in the analysis.

COMMENTS

The null-hypothesis is that no such thing as an aura exists. If this hypothesis were true, then the observations would usually be expected to be different from each other and different from the intention of the subject. The Kirlian photos would likewise be uncorrelated.

Stanford Research Institute recorded an RV success rate as high as 75%, [97] meaning that they chose the correct pattern 75% of the time from an essentially infinite number of alternatives. Applied to the current experiment, that would mean that each observer selected the correct aura pattern three quarters of the time, corresponding to two of the three observers agreeing on most observations. Naturally, with less experience (skill) we can't expect our rates to be that high, but it is possible.

PERSONAL OBSERVATIONS

Many times, I have asked someone to do something with his or her aura and then observed the result, using my etheric vision. In most cases, I have been at least a little surprised by what I perceived. For example, I have asked several people to reach out to a tree. In some cases, the person reached out as if with a hand. Some used a pipe-like appendage from the solar plexus or abdomen. Some reached out from his or her forehead. In each case, I have described what I saw and asked for an explanation. The person was usually surprised because he or she thought that they only imagined doing it.

In one case, I asked a man to change the shape of his aura in some arbitrary way. He changed it into a mathematical shape. I've never seen anything else quite like that. In fact, I wouldn't even have thought it was possible, if I hadn't seen it with my own etheric vision.

Now that we have (hopefully) established that auras exist, we have one point of reference on the etheric LOR. We know at least one type of object that

---

[97] Targ, Russell, and Katra, Jane, Ph.D. (1999) *Miracles of Mind: Exploring Non-Local Consciousness and Spiritual Healing* (p. 59). Novato, CA: New World Library.

exists on the etheric LOR. What about energy? Can we develop similar evidence of etheric energy? Let's see.

Chapter 22: Experiments with Etheric Energy 1

This experiment relates to something with which I have a love/hate relationship—a microwave oven. I wouldn't want to live without one, but I don't want to be near it when it's operating. I'll explain more about this below in my "Personal Observations."

Note that this experiment introduces an interesting possibility. If Kirlian photography works as I expect, we have the physical means to influence the etheric LOR and the physical means to record a characteristic of the effect. For example, we can use a microwave source to introduce energy on the etheric LOR and the Kirlian technique records part of the effect on an aura. This is a step toward instrumentation.

HYPOTHESIS

A microwave source causes energy to flow on the etheric LOR.

EXPERIMENTAL PROCEDURE

Rig a variable power microwave source to aim at the target area, but so that any electromagnetic (EM) leakage is constant—either minimal or zero—in the target area, not more than five feet from the source. No wood or paper is present between the source and the target.

Set up a meter at the target location to measure the leaked EM.

The subject is positioned beside the meter.

Kirlian photographic equipment is set up to record the subject's aura at selected times.

Three observers are located at least another ten feet away from the microwave source. They will use etheric vision to observe the subject's aura and the surrounding area on the etheric LOR. If using the AWIN system, remind the observers to try multiple vibration rates to gain maximal clarity.

Observations (including Kirlian photographs) are taken when the microwave source is off, at half-power (e.g. 1000W) and at full power (e.g. 2000W).

The observers take notes, then compare their notes with each other and with the Kirlian images.

METHOD 1

The subject does not need to have any skills, and makes no attempt to manipulate his/her aura.

The microwave source produces EM at least at a 2000W power level. The EM is aimed at the target, but shielding is used to reflect or deflect the EM so that no more than a tiny, constant amount reaches the target. If a more powerful source is convenient, I recommend its use. The more power, the more obvious are the effects

When the observers are ready, a Kirlian photograph is taken, the EM level is recorded and the observers make note of what they perceive with their etheric vision. The microwave source is turned on at half-power and the photo and observations are repeated. The source is turned to its maximum power and the photo and observations are repeated.

Without discussion, the observers take notes on their observations, including drawings and feelings. They compare their notes with each other and with the Kirlian photos.

METHOD 2

The subject is a house plant. All other elements of the experiment are the same. If the observers have difficulty perceiving the aura of the plant, just use the Kirlian photos.

METHOD 3

Take one leaf from the plant or from a tree. Repeat the experiment as in Method 2.

METHOD 4

Cut the leaf in half. Repeat the experiment as in Method 3.

COMMENTS

The null-hypothesis is that the subject's aura will not change at the different power levels, and no energy will be observed.

If there is concern about bias due to the sequence of power levels, add white noise loud enough to obscure the sound of the microwave source, and change the order. E.g. the power level at the three observations could be randomly selected between the three levels. Although this is a good idea for the human observers, the Kirlian images will not be affected by any physical cues, such as the sound of the equipment being turned on and off.

As with the previous experiment, the observers may use the AWIN techniques or RV. If RV is used, the observers may need to be separated so they do not disturb each other, and PCs can be used to coordinate their observations using something like instant messaging.

I have found that sufficient wood or paper blocks etheric LOR energy, so none should be present between the source and the target. After you have successfully demonstrated the ability to perceive the etheric energy created by the microwave source, you might choose to introduce a solid sheet of two-inch thick wood, and see what the observers perceive.

When the leaf is cut in half, carefully observe its aura and compare to the Kirlian photo. You may find the result surprising.

PERSONAL OBSERVATIONS

When a microwave oven is turned on, I see a cone of etheric energy. The cone extends downward on all sides of the oven, apparently corresponding to the beam of EM it generates. That etheric energy seems to disrupt the human aura, and I can feel it in my etheric body without using my etheric vision, so I try to stay out of the cone. I've noticed that some physical things (e.g. sufficiently thick paper or wood) block the etheric energy, and I think that the energy obeys the inverse square law.

Once, I was hiking in the Colorado Rockies with another sensitive person. We both felt quite uncomfortable at one point, so we stopped to look around. A microwave antenna was about fifty yards away, and we were nearly at 90° to its angle of emission. The etheric energy was so strong that we could see it damaging our auras. We got out of there quickly.

Chapter 23: Experiments with Etheric Energy 2

(Hopefully) we have seen that energy can exist on the etheric LOR. I chose microwaves, a physical source, to make it easier to study. The interaction between auras and the etheric energy was essentially incidental—an interesting sidebar—but I felt that it would make the observation of the etheric energy a little easier, since we had already established the existence of auras in a previous experiment, and you had experience observing them.

Most published studies on the nature of nonphysical LORs have had one of two orientations. Either the orientation was religious or human-centered. The religious orientation tends to limit the investigation to characteristics that are either good or evil—just as one can view the physical LOR, looking only at what you consider to be good or evil. Research based on this sort of religious orientation could not have developed our computer technology or our limited space travel, for example. The human-centered orientation is similarly limited, focusing primarily on human perception and human capabilities. If we had limited our knowledge of the physical in similar ways, we would know nothing of inorganic chemistry, electrical power, and many other current fields of science.

In my experience, the nonphysical LORs are at least as interesting as the physical LOR, with many things that have nothing to do with religious principles or human abilities. A few other authors have made similar statements related to the richness and complexity of other LORs. One phenomenon that is commonly known is that of "power spots." As mentioned earlier, at a power spot, etheric energy can be seen coming up out of the ground. Sometimes, I have had difficulty perceiving power spot energy. Some types of etheric energy are more easily perceived than others. I'll say more about this under Personal Observations below.

Some power spots are city-sized and low intensity. Some are a few feet across and intense. They come in all sizes, intensities and flavors. Yes, flavors. Each power spot seems to emit its own unique "type" of energy. I mention this primarily as a teaser, and do not intend to go into it further here. Small power spots are everywhere. You probably have one or more within thirty feet as you read this. The trick is to find one that is intense enough to be noticeable, and small enough to view easily from the outside.

HYPOTHESIS

Power spots exist and can be perceived with etheric vision.

EXPERIMENTAL PROCEDURE

RV, etheric vision, and/or astral travel is used to locate a nearby power spot of sufficient intensity to be interesting. The power spot is between 1 and 100 meters across at the base. Nine other locations are selected, some of which having no significant power spot, as far as can be perceived.

Three observers set up to observe the ten spots in random order. If using the AWIN system, remind the observers to try multiple vibration frequencies to attain maximal clarity.

Simultaneous observations are made.

The observers take notes, including drawings and feelings.

They compare notes.

METHOD

If etheric vision is used to observe the spots, the observers should set up within a few meters of its edge. Observers using RV can set up at whatever distance is convenient.

The details of one power spot vary widely from those of another. The observers compare their perceptions.

COMMENTS

Most well-known power spots are too large to be useful for these purposes, and their intensity is probably insufficient.

PERSONAL OBSERVATIONS

I have viewed and experienced many power spots in various areas of the world, some physically and some while astral traveling.

The first power spot I found was in Colorado, as previously mentioned. It was about fifty feet in diameter and oval at the base. The energy came out of the ground as a beam going straight up. I visited that power spot many times, both physically and in astral travel.

The Buddha is reputed to have gained enlightenment under the Bodi Tree in India. While visiting there on vacation, I found two power spots under that tree. One was directly under the tree while the other was nearby. Each was quite different from the other.

Carlos Casteneda wrote that it is possible to create an artificial power spot, but he did not include any details. On that same vacation in India, I saw an artificial power spot in a recently built temple. How did I know it was artificial? In my experience, a power spot has a reservoir out of which its energy flows. That reservoir has holes in it. As with all natural phenomena, the holes in the reservoir are irregular. But in this case, the holes were exactly regular. That power spot was definitely created to serve the temple.

Chapter 24: Life on Nonphysical LORs

I think of humans as multi-level organisms. In my experience, a human has active and important parts on all LORs. Animals and plants have different structures, but as we have seen, even plants have auras. I.e. they are also multi-level organisms. What about other life on the etheric LOR?

I have found that there is separate life on the etheric LOR. In fact, it abounds. I have no idea how many varieties of life there are. Over 100 years ago, C. W. Leadbeater wrote *The Astral Plane: Its Scenery, Inhabitants and Phenomena* [98] in which he described many things, mostly related to life. I suspect that his summarization of the various types of life on the etheric and astral planes will prove to be incomplete. As with many books on the subject, *The Astral Plane: Its Scenery, Inhabitants and Phenomena* has a religious flavor, but he made an effort to be scientific in his orientation, so the book is well worth reading.

Although I have read other books that might be useful in this investigation, they have been focused on one of the two perspectives I've mentioned before—either religion or human potentials—and I am concerned that researchers would be sidetracked by one of those perspectives, as so many have in the past.

---

[98] Leadbeater, C. W. (1895) *The Astral Plane: Its Scenery, Inhabitants and Phenomena*. London, UK: The Theosophical Publishing House. Can be viewed online
http://blavatskyarchives.com/theosophypdfs/leadbeater_the_astral_plane_its_scenery_inhabitants_1895.pdf

As for religion, yes, you will find things on the etheric and astral LORs that are easily thought of in terms of traditional religions—at this stage of our understanding. What I mean is that with more knowledge, things become less magical and more "normal." The nature of the thing doesn't change, but we become more able to understand it and begin to think of it as normal. For example, suppose you find a native in the deepest jungle who has never seen any modern technology. He would consider an SUV to be a monster. A digital camera would be an aspect of God or the Devil. In general, he would view our everyday mechanisms as magical, often related to God and to good or evil—just as most people view what they find on the etheric and astral LORs.

Is a tiger evil? Most people don't think so. Most people see a tiger as a superb predator, beautiful and graceful—but that doesn't mean they want to walk up to one in the jungle. Is a cow different from a tiger? In most respects, no, but we are educated to know the important differences, and we fear the tiger while we want to pet the cow. If we didn't understand either the cow or the tiger, we might categorize them as gods or devils.

What if we find life on other LORs that is more advanced than our own? What if it's more intelligent than humans? What if it's older and more powerful? As I said, our understanding of something doesn't change that thing, but it does change how we can relate to it. My opinion is that a modern scientific approach can be applied to understanding life on all LORs. In fact, I think this is an important step for humankind.

I do not wish to suggest any specific experiments for the investigation of life on nonphysical LORs. I am confident that as we become more aware of other

LORs, scientists will study the nature of life, just as they have on the physical LOR.

Chapter 25: Energy from Astrological Transits

In the first astrology experiment, evidence was found for the correlation between birth sun-signs and personalities. In subsequent experiments, evidence was gathered for the existence of auras, transit influences, and etheric energy. The present experiment ties these principles together, and observers begin to investigate the causal relationship between astronomical phenomena and the effects of transits.

To accomplish this, methods are combined from previous experiments. In the aura and energy experiments, you gained experience observing auras and etheric energy. In the transit experiments, you learned a crude method for estimating the intensity and strength of a transit. The same estimation method will be used here to select subjects who will be having major transits in the near future. The observation techniques will be used to characterize the subjects' auras before and during those major transits.

HYPOTHESIS

The influence of a major transit can be perceived as energy in the aura of the person having the transit.

EXPERIMENTAL PROCEDURE

A pool of subjects is selected, half of whom will have major transits within the next two months. The ones with major transits are placed in the "primary" group and those without significant transits are placed in the control group.

Three observers observe each of the subjects before the transit begins and during the transit. If using the AWIN system, remind the observers to try multiple vibration frequencies to attain maximal clarity.

Kirlian photographs are taken before and during the transits at the same times as the observations.

The observations are collected and compared with each other and with the photos.

METHOD

The observations are collected and compared with each other and with the photos. This experiment uses a double-blind method with a control group. The experimenter

[1] Determines which subjects will have major transits,

[2] Secretly distributes the subjects between the primary and the control groups, and

[3] Schedules the observations.

The observers do not know which subjects are having major transits or when, until all observations have been concluded. All observations of subjects in the control group are made at times when the subjects have not had any significant transits for at least two weeks. The first observation on each primary subject is scheduled about one week before a major transit begins and at least two weeks after any other significant transit has ended. The second observation is scheduled for one week after the beginning of the major transit. Only minor transits are ignored.

The subjects are not told that you are conducting an experiment related to astrology, since some of them might discuss their transits with the observers, potentially biasing the results.

The observers use either etheric vision or RV techniques. A combination of these methods would be preferable (i.e. some observers using each method).

Although I recommend the Kirlian photography, I do not know if the transit energy will show up on the photos. I hope that it does, as that would enhance this experiment, allowing mechanical comparison with the RV and etheric vision observations.

COMMENTS

Ideally, all observers do not know which subjects are having transits, but the experimenter may also be one of the observers. This might introduce some bias, but probably not.

I have left many details to be decided by the experimenter, based on experience with the previous experiments. Besides, you may have found ways to improve on my experimental designs. If you have, I hope you published them.

As suggested above, I think you will find in this experiment, a starting point for the study of the causal relationship between astronomical phenomena and astrological influences. The buildup of energy in the aura coincides with the presence of the transit. On further study, you will find that when the energy is absent, no significant effects are found from the transit. And you will see that the person can drain off the energy rather than allowing it to build up. Thus, the effects of the transit do not coincide with the transit; they coincide with the presence of energy from the transit. This rules out Carl Jung's theory of synchronicity. So where does the energy come from? What is the nature of the energy? Why is it predicted by angles between natal and current planetary positions?

PERSONAL OBSERVATIONS

In my own aura, I have observed a golden glow surrounding and somewhat penetrating my aura while I was having a major positive transit.

During certain types of negative transits, I have observed a vibration in my aura, as with a discordant combination of musical notes. Each transiting planet brings its own feeling in and around the aura, with negative transits feeling and looking different from positive ones.

As I have mentioned, with some knowledge, it is possible to channel the transit energy rather than allowing it to build up and discharge in an uncontrolled manner. In my experience, transit energy can be quite useful, even from negative transits.

CONCLUSIONS

We have arrived at an interesting place on this journey. Allow me to backup for just a moment to put this in perspective. People have been studying the planets and stars for tens of thousands of years. Long ago, they began to recognize patterns and correspondences that we call astrology. Records of magicians date back at least as far as the Egyptian Pharaohs. Magic and astrology were integral parts of science until the time of the Inquisition. Today, those fields are not part of science. Those, like me, who study magic, soon find that astrology is directly related. Since the astrological correlations are so strong and have been known for so long, why haven't they been incorporated into mainstream science?

Of course, all the same factors inhibit the incorporation of astrology as they do magic—the Inquisition, the institutionalization of science without the study of nonphysical LORs, and the social momentum that maintains it. But there is more. Many years ago, someone told me that mainstream science would not accept astrology, no matter how true it may be, until a causal relationship can be shown. In the mid-1970's, a Qabalist told me about the effects of transit energy on the human aura, and the potential to manipulate that energy. At that time, I was unable to observe the phenomena myself, but with some coaching, I was able to use the energy in a major transit I was having.

As you can see, the transit energy evident in the current experiment has been known for a long time. This appears to be part of the causal relationship between transits and the corresponding predicted effects. I think it's clear that there is more to the causal relationship than can be observed in the human aura. I have not checked, but I expect further investigation to uncover the emission of energy from astronomical sources that is unique to each source. Based on the observed correlations, it's clear that the angle of interaction between such energies causes the equivalent of interference patterns that we see in EM, such as visible light. I have related to this pattern of interference using an analogy to a system of magnets that create a net magnetic field measurable at a specific point. I know that is a crude analogy, but as yet, we lack the detailed knowledge to fully understand the causal relationship represented by astrology.

So where do we go from here?

Chapter 26: Other Research Directions

The previous experiments are primarily focused on what I have termed phase 1 of etheric LOR research wherein the intention is limited to demonstrating the existence of a few things—astrological influences, auras (human and plant), etheric energy from microwave sources, natural etheric energy from power spots, and etheric life. As I see it, these experiments provide a starting point. The last experiment suggests a causal relationship that will be seen to have far reaching consequences. Earlier chapters have covered some of the implications of astrology for human health and well-being, but the implications go far beyond that. Eventually, scientists will find that the implications are much broader than astrology, and affect much of what we know as life on Earth.

Although the transit energy experiment borders on phase 2 (determining relationships and characterizing), we really need to investigate the etheric LOR in much more depth and detail. This chapter discusses some of my ideas for further research directions, but I know that other people will have their own ideas, many of which I'm sure will be better than mine. But before going into directions I recommend, allow me to describe some directions I think can slow down or prevent the development of scientific knowledge.

In this regard, I've already mentioned the problems of a religious perspective. Just as in the times of the Inquisition, an investigation based on beliefs is likely to verify and support those beliefs. Even a strong desire can bias a person to ignore important evidence or to avoid specific research.

I've also discussed the limitations of studying human potentials at this time. When we have a better foundation in knowledge of nonphysical LORs, the study of human potentials will be more effective—in my opinion.

But there is another prevalent sidetrack that prevents people from gaining scientific knowledge of nonphysical LORs. This is the main attitude and approach of serious Qabalists, and is expressed in the writings of Carlos Castaneda—the search for power and/or advancement. I'm not saying that this is a bad attitude on an individual basis. What I'm saying is that people (like me) tend to dive deep and learn things that can be used for power and for advancement. I think William G. Gray expressed the attitude fairly well in his *Inner Traditions of Magic* when he said, "Who now *knows* what to think, *dares* to think it, *wills* to do something with it, and has the sense to *keep silent* about experiences of Inner Life? Such was demanded of Magical Initiates in olden times, and we have not yet outdated the necessity for those Four Axioms." [99] (By Inner Life, Gray meant nonphysical LORs.) Interestingly enough, the forward of this book, written by Israel Regardie begins, "Had this book been written five hundred years ago, William Gray would have been seized by the Inquisition and burned at the stake."

I'm not sure that it's valid to call it a trap. If you want to travel to a distant mountain, you probably choose a convenient form of transportation rather than inventing a new type of engine for a vehicle. Analogously, the people attracted to study nonphysical LORs have typically wanted to travel to the

---

[99] Gray, William G. (1970) *Inner Traditions of Magic* (p. ix). New York, NY: Samuel Weiser Inc. Can be read online at http://www.oldways.org/documents/ceremonial/gray/gray_inner_traditions_of_magic.pdf

mountain. The problem is even more difficult, though. Even those who just want to develop a new engine (e.g. gain scientific knowledge of nonphysical LORs) become so excited about the distant mountain that they forget their original purpose and head out for the mountain (i.e. the pursuit of power and/or advancement).

This is, however, similar to a trap. Many people, having started toward the mountain, can think of little else. Perhaps it is similar with a man who begins to run for exercise and becomes addicted to the endorphins—running becomes an obsession. In the study of nonphysical LORs, a scientist may likewise become obsessed with the pursuit of power or advancement. I know I did.

Even so, going deep is not always the wrong thing to do. Scientific investigation is seldom a linear process wherein the steps are clear, laying a solid foundation before going further. In the terms that I have used here, I conceive of the discovery process as a misty mixture of phase 1, 2, and 3 in that even while verifying the existence of one thing, we may be studying relationships with other things and refining measurements.

Let's consider another pitfall. Remember Francis Bacon's statement about the discovery of truth—basically to take one step at a time rather than jumping to conclusions. Much has been written about the nature of nonphysical LORs. All too often, it's tempting to accept another person's conclusion and then figure out how your knowledge fits that conclusion, especially if you prefer it. Although it's important to know what conclusions others have drawn concerning nonphysical LORs, as we gather data, that data should be the basis of our conclusions. Quantum Physics and recent parapsychology experiments have developed some unintuitive results. And just how intuitive is the special theory of relativity? In other words, I think it is important to follow Bacon's advice, no matter where the evidence leads.

Now I'd like to suggest a few additional investigations to develop our scientific knowledge of nonphysical LORs. I won't lay them out as detailed experiments, but I think the methods used in the previous experiments are suggestive of how to structure the experiments, make observations and analyze the results. I think it's most appropriate to express them as questions.

What is the basic unit of objects on the etheric LOR? In physical matter, we generally agree that things called atoms are the building blocks. We have gone further and talk about atoms being composed of electrons, neutrons and protons, but I would be happy if we knew the equivalent of atoms for etheric objects.

Is there something in etheric objects equivalent to chemistry? I have observed qualitative differences in etheric objects. I suspect the answer relates to energy, but I don't know.

What is the relationship between objects and energy on the etheric LOR? On the physical LOR, we know that matter can be converted to energy and energy can be converted to matter, but the means of conversion from one to the other is quite limited. My opinion is that on the etheric LOR, objects can be formed from energy and converted back to energy with fewer restrictions. But that's just my opinion. What does the evidence show?

Where does power spot energy come from? In my observations, it seems as if the energy may actually be coming from the astral LOR to accumulate in the reservoir of the power spot. Am I right or wrong about that?

How can the aura be characterized? What is it made of? Does it have inertia? How does it change shape or size? I have observed a surface on auras, and that characteristics of that surface have effects. As a guess, I would say that an aura is relatively complex.

What is the relationship between a person and his (her) aura? What happens to the aura when the person sleeps? If you amputate a hand, the aura still has that missing hand—or so I've read. A Kirlian photo of a leaf that's been cut in half still shows the aura of the full leaf—or so I've read. And as we've seen, the person can change the shape, size and location of his/her aura. I have observed many psychological (and some physical) reactions related to a person's aura.

Many people have observed "chakras." What are they? People have reported them as "energy centers." Are they part of the aura?

What happens when two auras interact? How many auras can overlap?

What are the differences between human auras and plant or animal auras? I suggest similar experiments on a variety of species and types of plants.

Why do different power spots have different "flavored" energy? Is it the nature of the source of the energy? Is it the nature of the power spot's reservoir? Or is it something entirely different?

What is the nature of the things commonly called "elementals?" E.g. *something* apparently alive and conscious to some degree is associated with a mountain. It is commonly called an "earth elemental." Other elementals are associated with bodies of water, fires, and storms. Many people have observed these things.

How can we measure etheric energy? The experiments with microwave sources are suggestive. Perhaps we can measure etheric energy relative to the intensity of a microwave source. Is the effect of the microwave source linear on the etheric? Is it inversely proportional to the distance from the source?

For example, what if we could measure the intensity of astrological energy? A few opinions about measurements. Humans originally measured relative to their own bodies—height in terms of hands or a man's height, distance in terms of feet or strides, etc. Eventually, the measurements were standardized, but that required much advancement. I think we're in the initial stages of measurement on the nonphysical LORs, actually just beginning to define what should be measured. Relationships are generally needed to define units and measurements. We have found a few relationships. Let's use them.

We've been talking about the etheric LOR. What about the astral LOR? What are the differences between etheric LOR objects and astral LOR objects? What are the differences between etheric energy and astral energy?

In fact, when you are able to perceive the etheric and lower astral LORs, you will realize that the nonphysical LORs are continuous. I think of the LORs like the depths of the ocean. Yup, another analogy. Think of the physical LOR as corresponding with the bottom of the ocean, and the levels of water above the bottom correspond to the etheric, astral and other LORs. The levels of the ocean can be characterized in terms of ecological systems. I see the LORs as continuous, like those of the ocean, and I see things moving up and down the LORs much as fish swim up and down in the ocean. I realize that this complicates the investigation, but doesn't it also make it more interesting?

Section VIII: Now The Door Can Be Opened

Chapter 27: Summary

We are at the end of our journey together—or is it the beginning? Time will tell. Where have we been on this journey? We began with a strange young man who had the illusion of knowing something was about to happen. Wasn't that what you thought? But over the next few years, that young man, oh, yeah, that was me, I got external validation that was even more convincing than my personal experience. So why, if those experiences were true, don't mainstream scientists investigate?

We learned that science *did* include some knowledge of nonphysical LORs until the Inquisition threatened the life of any scientist who dared to study it. Then, for four hundred years modern science developed under that flaming torch, and became institutionalized, developing sufficient social momentum to perpetuate itself, warts and all. And today, that social momentum is carried forth by parents, teachers, scientists, administrators, factory workers, and just about anyone else you can think of, in Western society—except the government. Even as the rulers during the Inquisition used magi and astrologers, [100] our modern governments secretly develop and use nonphysical LOR technology. [101]

---

[100] Professor Teofilo F. Ruiz. Professor at the University of California at Los Angeles. Terror of History: Mystics, Heretics, and Witches in the Western Tradition course located at http://www.thegreatcourses.com/courses/terror-of-history-mystics-heretics-and-witches-in-the-western-tradition.html

[101] Targ, Russell, and Katra, Jane, Ph.D. (1999) *Miracles of Mind: Exploring Non-Local Consciousness and Spiritual Healing* (p. 38). Novato, CA: New World Library.

Nonphysical LORs are not as obvious as gravity, well, maybe almost that obvious, but not quite. Astrology is pretty obvious, as it is studied and used all over the world—it's just not condoned by mainstream science. Even something as obvious as astrology can be ignored if you don't look at the evidence—like Carl Sagan. [102] All you have to do is believe what you've been taught.

Other, slightly less obvious aspects of nonphysical LORs, have been openly studied and used before the Inquisition began and within fifteen years after the torch was taken away from the Inquisition. Societies like the Golden Dawn teach those who are ready to learn, how to gain power and advancement. Parapsychologists have been demonstrating nonphysical-based human potentials for over seventy years in scientific experiments, but their results have not been incorporated into mainstream science.

In recent years, we have developed methods that can be used to make direct observations of events on nonphysical LORs, most prevalent among them being the RV systems. The AWIN system is a new version of something much older—the Qabalah. We have seen that although far from infallible, RV and AWIN techniques can be used effectively to make scientific observations. Since observations are the key to the scientific method, [103] scientists now have the means to investigate other LORs.

---

[102] Sagan, Carl. (1996) *The Demon-Haunted World: Science as a Candle in the Dark* (p. 304). New York, NY: Random House.
[103] Bacon, Francis. *Advancement of Learning and Novum Organum: With A Special Introduction by James Edward Creighton, PhD* (p. 316). Published in New York (1900) by the Colonial Press. Can be viewed online at http://www.biodiversitylibrary.org/item/59907#page/396/mode/1up

But even when you have the means to do the research, why would you go against the social momentum that so powerfully perpetuates the misinformation about nonphysical LORs? We already know that the downside is ostracism, loss of respect, and potential loss of funding. What would the upside be? How about the pursuit of truth? For most people, that is not enough reason to endanger your career. So why do it? Why move science forward in the face of so much resistance?

First, the resistance is waning. More and more scientists today are interested in nonphysical reality. Movies and TV shows about nonphysical LORs are becoming more popular, such as The Craft, Charmed, Ghost Whisperer and others. The Oprah Winfrey show featured *The Secret* by Rhonda Byrne. Police commonly use psychics to solve difficult crimes. These and other indicators show a general slowing of the social momentum, and a willingness to consider the truth that we can discover in all LORs.

Second, the benefit to humankind would be tremendous. The benefits in health alone would be great. I can see why some doctors might resist the changes, but what's wrong with more fields of specialization within medicine and psychology? Wouldn't you rather live in the world described in the fictional chapter, where people stay healthier and happier?

If those are not sufficient reasons to move forward, add fame and fortune. Now that people know more about the truth and how misconceptions have been perpetuated about nonphysical LORs, more serious and intelligent people will begin to learn and to perform experiments. This is a tremendous opportunity, larger and more significant than the development of physical-LOR science. The people who investigate will answer many important questions, and go down in history as the discoverers of laws of nature. The technological opportunities are also considerable. Some people are going to make a lot of money developing them. Are you sure you don't want to be among these pioneers?

Appendix A: Sun-Sign Related Personality Descriptions

This appendix provides initial descriptions that can be used in the sun-sign experimental procedure.

1 Aries
2 Taurus
3 Gemini
4 Cancer
5 Leo
6 Virgo
7 Libra
8 Scorpio
9 Sagittarius
10 Capricorn
11 Aquarius
12 Pisces

Table 4: Astrological Signs

PERSONALITY TYPE 1

As a child, they were noisy and full of energy. Always independent, they have to live life their own way.

They have courage and strong will. They love change, and don't like routine.

Their first ambition is to be first—they love to compete. They enjoy winning.

They are frequently original in their efforts to be first, and, when not original, sure to be novel. When original, they are pioneers, inventors, great thinkers. When only novel, their ideas may not be sound or practical.

They can be self-centered; if so, they can be arrogant, conceited, self-pitying, self-assertive, aggressive, blunt, impatient.

They can be good leaders—energetic, setting the direction and charging forward.

They're self-starters but apt to lose interest if the pace slows down or things become complicated. They enjoy starting things, but have less interest in finishing them.

MALE

He is quick-thinking and aggressive, with a high vitality. He has a quick temper, but seldom holds a grudge, and doesn't think anyone else should, either. He's always busy doing something. He doesn't want to wait for anything or anyone, especially lazy or slow people.

He may be egotistical and amoral. If so, he is probably also self-centered, destructive, ruthless, guiltless, and enjoys intimidating people.

FEMALE

If she's not too self-centered, her high energy draws people to her. She tends to be enthusiastic and extroverted. She prefers people who are direct and get to the point quickly.

She is likely to open the door for herself, pull out her own chair, etc. She is strong-willed and more independent than most women.

If she is a professional woman, she is quite focused on her career, talking about it often. Her self-esteem is tied to her career.

She may be self-centered and egotistical. If so, she is at best a "pushy broad," being bossy, controlling, arrogant and insensitive.

PERSONALITY TYPE 2

These people are slow, steady and stubborn. Their reflexes are slow and they cannot be pushed. In dealing with them, indirect action is best.

They tend to be disciplined, orderly and reliable. They are usually conservative in outlook and attitude.

It takes time for them to adjust to new ideas. They get accustomed to new concepts gradually. They get flustered and upset if they are not given time to assimilate facts slowly.

It takes a great deal to anger them, but when they do get angry, you can't argue with them until they cool down. They can hold a grudge.

Positive qualities include singleness of purpose, loyalty, and stick-to-itiveness.

Anyone who is going to get along with this type of person must understand that to them, cooperation doesn't mean doing things together—it means doing things peacefully, in a friendly manner, even if the things are done separately.

MALE

He is possessive and probably jealous.

If not lazy, he is strong, steady, sincere and practical. He has the kind of perseverance that gets things done, but not necessarily quickly.

If lazy, he probably also has a potbelly, and loves to eat. In this case, he is probably greedy, happier to take than to give. He wouldn't think of changing his routine.

FEMALE

She is typified as the earth mother; slow, steady and strong. She is quite feminine.

Mothers with this personality type are quite possessive. She finds it difficult to let their children go, even when they are adults.

She may be overweight. For her, stability and safety are key concerns.

PERSONALITY TYPE 3

They are versatile and adaptable, but not noted for concentration and persistence.

They are mental in orientation and can be neurotic. If you wish to influence them, you must appeal to them through logic and reason.

They dramatize things that happen to them and are apt to let their imaginations carry them away.

They like to try new things, and generally do what they say.

They desire to be themselves, to think for themselves and to do for themselves.

They are usually liberals. Many of them like to gamble. While young, they like free love, breaking school rules and talking back to cops.

They tend to be witty, with a good sense of humor. This makes them endearing to many people. They can sell ideas to anyone. They are generally adverse to close and binding ties. The more liberty they get in marriage, the happier they are.

They like a vocation in which they can move about and mingle with people, rather than sitting at a desk for eight hours.

MALE

If positive, he is charming and enthusiastic. He is quite intelligent and is intellectually oriented, typically moving in multiple directions at once. He tends to be the life-of-the-party.

He is fun-loving.

If negative, he is cold, detached, inconsiderate and undependable. Such a man tends to change his mind frequently, and is a shallow pleasure seeker.

FEMALE

She talks a lot, and even if she's quite intelligent, sometimes she drives you crazy with her babbling.

She tends to be critical.

She's usually late getting to an appointment, party, meeting, etc.

She may be superficial and shallow. If she is, then her feelings are probably also shallow, and she will not give of herself in a relationship.

PERSONALITY TYPE 4

They are passive, receptive and emotional in nature, capable of much sustaining and nurturing. They tend to live in their feelings and affections. Not forthright and direct in action, they often sidestep or cloud issues. They respond to life through their feelings, and you can't reason with them when they're emotionally disturbed.

They usually have a strong attachment to their home and mother.

They are not likely to enjoy exercise.

Their lives aim primarily at security—emotional, material and domestic. Capable of great self-sufficiency, or of being a clinging vine, they must always be secure. Their sense of responsibility toward another's money, security, etc., is as deep as if it were their own. They pay their debts and expect others to do the same.

They are very good as home-makers. The maternal-paternal instinct is powerful, and these people, male or female, will go to great lengths to protect, defend and improve their homes, mates and children.

MALE

He can be quite sympathetic, especially for the underdog.

If he is positive and secure, he is imaginative and creative in artistic ways. He is tenacious when he sets goals for himself. He is generous and caring.

If he is insecure, he is emotional and moody, often suspicious and over-sensitive. His communication is indirect, especially if he feels wronged, and he is easily offended.

FEMALE

She is moody, but capable of much mothering.

If she is positive, she tends to offer emotional support and assistance to others when they need it.

If she is negative, she is concerned with her own needs, rather than those of others, and she is quite insecure.

PERSONALITY TYPE 5

They are natural leaders and have great courage. They are self-confident and loyal, but they can be overbearing. They have much pride. They're bold, creative, dramatic and strong.

The foundation of their considerable energy, vitality and charm is their desire for both public acclaim and self-approval.

They like being the center of attention. They are not genuinely introspective, but they are willing to change to improve their effect on others. They are basically honest, and as a result, as they act, so they become. They care about what people think of them.

At their best, they are earnest, sincere and eager to please, and willing to go to much effort to do so.

If they are shallow, they can be vain, pompous and proud.

MALE

He is easily flattered.

FEMALE

She makes an effort to look her best, whether that be with clothes, her slim figure, makeup, money, etc.

PERSONALITY TYPE 6

They collect, analyze and correlate facts to improve themselves and the world. They are responsible, desiring order and perfection. They are among the most organized of people.

They have an inferiority complex. Taking care of endless details and routine jobs is where they function best. There is a cautiousness and often a selfishness that they probably don't recognize. It is difficult for them to communicate and articulate where their inner thoughts and feelings are concerned.

Home is extremely important to them for they feel more secure there than in any other surrounding. They are not gregarious. They are basically shy. They have jittery nerves and a highly geared nervous system.

They identify themselves with their work and are quite willing to lose themselves in it.

Their greatest fault is being too critical. Their analyzing minds can cause them to degenerate into faultfinding, criticism and irritability.

They are practical and down to earth in their approach to life. They dislike anything crude or coarse. There is a sweetness and a lack of aggressiveness in their nature that gives a great attractiveness. They make few enemies and have many friends due to a quiet, gentle manner.

MALE

He is well-organized and industrious. Many men with this type of personality only work and sleep—nothing else interests him—except love.

His accomplishments are very important to him. He considers efficiency to be goodness.

He may hold a grudge indefinitely.

FEMALE

She believes in moral principles and privacy (unless she talks incessantly). She is emotional, with delicate feelings.

PERSONALITY TYPE 7

Relationships are very important to them. They want to be all things to all people, as long as their deep-seated values are not violated. They usually prefer to be liked by everyone rather than take a stand on an issue. "Peace at any price" is their motto. However, when integrity is involved, they draw the line and set their feet.

It is difficult for them to be generous. They are secretive about finances and personal matters.

They make good diplomats because they're friendly, outgoing and interested in living life to the fullest. They are social and do well in groups because they are tactful and conscientious, have a strong sense of justice and the will to do good.

They tend to look at both sides of an issue.

MALE

His charm and sense of humor attract people at a party. He likes to entertain, much as an actor does.

He likes music, whether he plays an instrument or not.

He is sensuous. If he is well-adjusted, this manifests as a love of good food. If not, he is probably a fickle playboy.

He may be articulate, with the ability to synthesize diverse factors—which is beneficial in business.

FEMALE

She wants attention and needs to feel appreciated. She's feminine and sensual. She tends to be creative and can be artistic. She finds it difficult to ask for what she needs.

PERSONALITY TYPE 8

There are two different expressions of this personality type—either angelic or nasty. If angelic, moral attitudes and actions are of the highest order. If nasty, moral attitudes and actions are of a low order. In either case, they are intense.

They are mysterious, in that much is hidden beneath the surface. In most relationships, they command respect by their strength.

If of the angelic expression, they have tremendous strength and power, especially when in a position to serve other people in some significant way, e.g. as a doctor or nurse.

If of the nasty expression, they can be vindictive, jealous and resentful, especially among family. In this case, the tongue can be deadly, cutting deeply with only a few words.

They can be quite stubborn. They have strong passions and tremendous pride.

MALE

He has great insight into people. He may care deeply about people, in which case, he is kind, compassionate and concerned. He can overcome any obstacle and enjoy the challenge.

If he doesn't care, he uses his power and insight to play mind games, and usually wins. Such a person can be a sociopath, self-serving and proud of his accomplishments, no matter how much destruction he caused along the way.

FEMALE

She is complex. She may be kind, compassionate, psychic and intuitive, with something about her that makes people tell her their deepest problems. Or she may be crafty, manipulative and self-serving.

PERSONALITY TYPE 9

Most people with this type of personality are friendly, outgoing, optimistic and extroverted. They probably love sports and gambling. They are willing to take a chance on just about anything.

They are independent and freedom is important to them. They can't be pinned down.

They don't care about tradition or the past—they are interested in the future and in change. They have many interests.

They may love travel and sports, and they may be interested in spiritual philosophies.

They are good at sales and promotion.

They tend to procrastinate.

They tend to be able to hit a person's weak point, so they can have trouble with people if lacking in tact.

MALE

He enjoys life and makes people laugh. In fact, he can be quite entertaining. He is impulsive.

FEMALE

She has a positive attitude and most people like her and want to be around her.

She probably has trouble being punctual.

She tends to be too masculine with men and may have trouble in relationships.

She may suffer from foot-in-mouth disease.

PERSONALITY TYPE 10

They have great strength and a strong sense of purpose. They may use this for good or for ill.

They want to be leaders, and probably think they know what is best for other people.

Money is important to them for the power and prestige it brings.

They tend to be patient, persistent, organized, efficient and practical. They are ambitious and willing to work hard for what they want. They intend their achievements to endure, and not just be fleeting.

They value tradition, home, mother and the past. They may love antiques.

They may be nicer outside the home than they are to their families.

MALE

He is a disciplinarian with his children.

He may be condescending and controlling.

FEMALE

She probably strives for money and prestige, whether by her own effort or by marriage. She is self-critical, and never achieves enough, even though she may become the CEO of a major corporation—or marry one.

She is emotional, but controls them with an iron grip. She may seem cold and uncaring.

PERSONALITY TYPE 11

They tend to live more in the future than in the past. They are outgoing and friendly. They seem to be self-confident. They cannot be pushed into anything they don't want to do. When interacting with people, they are more impersonal and detached than emotional.

They dislike restraints and have to go their own way. They are independent, idealistic, imaginative, creative and inventive. They are strong-willed and may have difficulties in marriage or in unions.

They may seem aloof and separate from other people. If so, they might say, "I love humankind. It's people I can't stand."

As a manager, they may be critical and demanding.

Probably has two types of friends—one type is conservative (traditional) and the other type is unconventional (bohemian).

They are capable of working well in organizational work or in big business. They may be capable of being good leaders and organizers.

MALE

He is quite involved in his work, probably in an emotional way.

If he is frenzied, he probably grins a lot and may be manipulative.

FEMALE

She is curious about many things, especially about the strange and unusual, whether it be social, animal, science, or any other topic.

She may be self-serving.

PERSONALITY TYPE 12

They are sensitive and their emotions are strong and deep. They are moody, introspective and hard to understand. They need to be alone and retreat from contact with the world in order to retain equilibrium. When they are well-balanced, they are capable of great achievements, but when unbalanced, they may attempt to escape through alcohol or drugs.

They are intuitive and often empathic, able to sense and react to others' emotions.

They have deeply hidden inner pride, but they are not particularly competitive.

They suffer from an inferiority complex and a sense of unworthiness. They never feel that they do enough, so they often overwork.

They have a deep love of music and can be excellent musicians if they pursue it.

MALE

If not self-destructive, he is spiritual in his attitude toward life. He is interested in forms of healing, such as medical, acupuncture, etc. He is also interested in alternative approaches such as yoga and Tai Chi.

If self-destructive, he is a pleasure seeker and dreamer. He wants what he can't have. He may dive into drugs and sex.

FEMALE

She seldom has a clear self-image, and often gets into relationships with the wrong people.

She is sensual. Some express this in positive ways, with perfume, scented baths and love-making.

If not self-destructive, she longs for the fairy tale life and is artistic in nature. She is compassionate and caring.

If self-destructive, she may live in drama. She may get lost in drugs, sex or excesses of food. She probably has many unpleasant experiences with the wrong kinds of people.

Such a woman is likely to be passive-aggressive. Good advice from anyone goes unnoticed.

Appendix B: Subject Surveys

PART 1: Subject Qualification Survey

I am (We are) running a focused experiment on personality types and what people think about each other. To participate, please answer the following ten (10) questions on a scale from 1 to 10, where 1 indicates no knowledge of the topic, and 10 indicates expert knowledge of the topic.

Name: ______________________________

1. __ The scientific method

2. __ Psychology

3. __ Human perception

4. __ Human personality types

5. __ Personalities of those people you know well

6. __ Statistical analysis methods

7. __ Quantum theory

8. __ Astrology

9. __ Aliens

10. __ Conspiracy theories

PART 2: Potential Target Identification Survey

Please provide the names and contact information for people you know well. These may be long-time friends or relatives, or they may be people you work closely with.

Your Name: ______________________________

Name: ________________________________

a. Contact information

i. Telephone: ______________

ii. Email: _________________

b. Your relationship to this person: ________________

c. How long have you known this person? ___________

d. How recent is your experience with this person?

i. __ Within the last few months

ii. __ Last year

iii. __ More than 1 year ago

NOTE: Provide the individual with multiple copies of this survey in order to secure names and contact information.

Appendix C: Surveys for Qualifying Targets

I am (We are) running a focused experiment on personality types and what people think about each other. To participate, please answer the following questions to the best of your knowledge. If you choose not to answer a question accurately, please leave it blank. Your age and birth date may be used to group participants, so please enter them accurately.

Name: ____

1. Age __________

2. Birth Date ___________

3. Please mark you cultural background

   a. __ Farming

   b. __ Blue-collar (e.g. factor workers, trucking, etc.)

   c. __ White collar (office workers)

   d. __ Other

4. Please select all of the following personality type characteristics that apply to you

a.	__ Conservative	g.	__ Supportive
b.	__ Possessive	h.	__ Hot tempered
c.	__ Liberal	i.	__ Even tempered
d.	__ Permissive	j.	__ Strong-willed
e.	__ Controlling	k.	__ Easy-going
f.	__ Giving/generous		

Appendix D: Meditation Instructions

This appendix describes how to use the AWIN meditation methods. Meditation is the foundation of the AWIN system. It is the starting point, and most people will use it often when they practice the AWIN techniques. Meditation is necessary because we have been taught to ignore our perceptions of the etheric and astral LORs and focus completely on the physical LOR. The point of meditation is to weaken the normal strangle hold of your conscious mind on your awareness. By using discipline, you will gain more control over your awareness without becoming confused. I.e. if you follow these recommendations, you will always know what is physical reality and what is not.

The methods provided here teach your mind new habits, allowing you to gain control over previously unconscious filtering. It is important that you develop these habits carefully. If they are developed haphazardly, your meditation will be ineffective, and it will be difficult or impossible to learn the AWIN techniques. Follow the **USAGE** section and you should have no problems.

The methods taught here consist of successive changes in the meditation technique. The first two steps teach your mind to 'let go' and allow you to relax. Many relaxation techniques are taught for many purposes. Feel free to combine other relaxation methods with this meditation, as long as they do not require lengthy exercises. After a few weeks, you should be able to go into meditation in less than five minutes, so a relaxation method that requires a half hour will not work for you.

The second step is to teach your mind that it is OK to relax more quickly and to be aware of something other than physical reality. You do this every night in your sleep, but you were taught that sleep is only imaginary and should be ignored. When you were little, you were aware of other LORs, but you've probably forgotten about it. Now it's time to cultivate what you forgot.

The third step is to teach your mind to relax quickly and that it is OK for your etheric body to act independently of your physical body. Again, you do this in your dreams, but your conscious mind finds it scary, freaky, and uncomfortable. It's really no big deal. Treat it with the same respect you use to drive a car, and you won't run into trouble.

For simplification in the rest of this technique, the content of the MP3 file or CD will simply be referred to as "the CD."

USAGE

Find a place where you can be comfortable and will not be disturbed. Too much noise will be distracting, so the place should also be reasonably quiet. You should not be too tired since you should not go to sleep. You will also find it easier if you are not emotionally upset. Get as comfortable as you can, but a sitting position is usually the most effective. Play the recording and do as it says. The recordings are exactly as written in the **METHOD** sections, and I recommend that you first read a method before you practice it using the recording. By so doing, your conscious mind will know exactly what to expect. It will not be surprised, and in fact it is likely to get bored. This is good for our present purposes.

CAUTION AND TIPS

[1] Although you shouldn't go to sleep, you may think you did, the first time you listen to the CD. Don't worry about it the first time, but try to be rested. A sitting position is recommended when using the CD.

[2] Best results can be achieved for most people when not too full or too hungry.

[3] Keep a journal of your meditation experience. In each journal entry, make a note of all things that might influence your meditation, including the following:

¤ Emotional state (ES)

¤ Mental state (MS)

¤ Physical state (PS)

- Time since last ate (TSA)
- Fullness of stomach (FS)
- Degree of restedness/alertness (R/A)
- Amount and intensity of pain (AIP)
- Degree of illness/health (DIL)

[4] Read through each meditation method before you listen to the CD. That way you know what will be on the CD and you can trust it to lead you safely.

[5] You may choose to meditate more than once per day. That is fine. It is best to practice meditation at least three times each week while you are learning to meditate. The AWIN techniques presented here can safely be practiced many times per week.

[6] Your etheric body normally coincides with your physical body. These meditation techniques teach you to become aware of your etheric body. You will learn to travel in your etheric body. This is just as natural as walking, and you often do so at night while you sleep.

NOTE: Always return to your physical body before coming out of levels. Although you would probably not be damaged by coming out of levels while your etheric body is away from your physical, it could cause problems and would most likely reduce the effectiveness of your techniques.

METHOD 1 [104]

This is a full meditation method, including the use of tension to increase relaxation. Although you may choose to use it from time to time, this method is intended to be used only when you are first learning to meditate.

The audio recording of this method is 27 minutes long.

After you have practiced this method for a week or two and at least several times, you should graduate to Method 2.

TRANSCRIPT OF METHOD 1

Make yourself comfortable. Direct your eyes upward at an angle of 45 degrees. Keep your head straight and focus on one spot.

Mentally repeat and visualize the number five. Five. Five. Five. Five. Five. This is level five, your normal waking level of mind.

---

[104] http://www.awinsystem.org/Handouts/MedMethod1.mp3

Close your eyes. Keep your eyes closed until I tell you to open them again.

Mentally repeat and visualize the number four. Four. Four. Four. Four. Four. You are now at level four.

To help you relax, I will direct your attention and help you relax each part of your body.

Now direct your attention to your forehead. Tense your forehead. Feel the tension build. Now slowly relax and release all tension in your forehead. Feel how relaxed it is.

Now direct your attention to your face. Tense your face. Feel the tension build. Now slowly relax and release all tension in your face. Feel how relaxed it is.

Now direct your attention to your neck. Tense your neck. Feel the tension build. Feel the organs within your throat. Now slowly relax and release all tension in your neck. Feel how relaxed it is.

Now direct your attention to your shoulders. Tense your shoulders. Feel the tension build. Now slowly relax and release all tension in your shoulders. Feel how relaxed they are.

Now direct your attention to your chest. Tense your chest. Feel the tension build. Now slowly relax and release all tension in your chest. Feel how relaxed it is.

Now direct your attention to organs within your chest area. Feel your heart beating and your lungs filling with air. Feel them functioning in a relaxed and healthy manner.

Now direct your attention to your back. Tense your back. Feel the tension build. Now slowly relax and release all tension in your back. Feel how relaxed it is.

Now direct your attention to your abdomen. Tense your abdomen. Feel the tension build. Now slowly relax and release all tension in your abdomen. Feel how relaxed it is.

Now direct your attention to the organs within your abdomen. Feel them functioning in a relaxed and healthy manner.

Now direct your attention to your buttocks. Tense your buttocks. Feel the tension build. Now slowly relax and release all tension in your buttocks. Feel how relaxed they are.

Now direct your attention to your thighs. Tense your thighs. Feel the tension build. Now slowly relax and release all tension in your thighs. Feel how relaxed they are.

Now direct your attention to your calves. Tense your calves. Feel the tension build. Now slowly relax and release all tension in your calves. Feel how relaxed they are.

Now direct your attention to your feet. Tense your feet. Feel the tension build. Now slowly relax and release all tension in your feet. Feel how relaxed they are.

Now direct your attention to your scalp. Feel as though a warm bath surrounds and interpenetrates your scalp. It calms and sooths. Feel this part of your body going to a deep state of relaxation. Feel the relaxation spreading to nearby parts of your body.

Now direct your attention to your forehead. Feel as though a warm bath surrounds and interpenetrates your forehead. It calms and sooths. Feel this part of your body going to a deep state of relaxation. Feel the relaxation spreading to nearby parts of your body.

Now direct your attention to your eyelids. Feel as though a warm bath surrounds and interpenetrates your eyelids. It calms and sooths. Feel this part of your body going to a deep state of relaxation. Feel the relaxation spreading to nearby parts of your body.

Now direct your attention to your face. Feel as though a warm bath surrounds and interpenetrates your face. It calms and sooths. Feel this part of your body going to a deep state of relaxation. Feel the relaxation spreading to nearby parts of your body.

Now direct your attention to your neck. Feel as though a warm bath surrounds and interpenetrates your neck. It calms and sooths. Feel this part of your body going to a deep state of relaxation. Feel the relaxation spreading to nearby parts of your body.

Now direct your attention to your shoulders. Feel as though a warm bath surrounds and interpenetrates your shoulders. It calms and sooths. Feel this part of your body going to a deep state of relaxation. Feel the relaxation spreading to nearby parts of your body.

Now direct your attention to your chest. Feel as though a warm bath surrounds and interpenetrates your chest. It calms and sooths. Feel this part of your body going to a deep state of relaxation. Feel the relaxation spreading to nearby parts of your body.

Now direct your attention to the organs within your chest area. Feel as though a warm bath surrounds and interpenetrates your torso. It calms and sooths. Feel this part of your body going to a deep state of relaxation. Feel the relaxation spreading to nearby parts of your body.

Now direct your attention to your back. Feel as though a warm bath surrounds and interpenetrates your back. It calms and sooths. Feel this part of your body going to a deep state of relaxation. Feel the relaxation spreading to nearby parts of your body.

Now direct your attention to your abdomen. Feel as though a warm bath surrounds and interpenetrates your abdomen. It calms and sooths. Feel this part of your body going to a deep state of relaxation. Feel the relaxation spreading to nearby parts of your body.

Now direct your attention to your buttocks. Feel as though a warm bath surrounds and interpenetrates your buttocks. It calms and sooths. Feel this part of your body going to a deep state of relaxation. Feel the relaxation spreading to nearby parts of your body.

Now direct your attention to your thighs. Feel as though a warm bath surrounds and interpenetrates your thighs. It calms and sooths. Feel this part of your body going to a deep state of relaxation. Feel the relaxation spreading to nearby parts of your body.

Now direct your attention to your knees. Feel as though a warm bath surrounds and interpenetrates your knees. It calms and sooths. Feel this part of your body going to a deep state of relaxation. Feel the relaxation spreading to nearby parts of your body.

Now direct your attention to your calves. Feel as though a warm bath surrounds and interpenetrates your calves. It calms and sooths. Feel this part of your body going to a deep state of relaxation. Feel the relaxation spreading to nearby parts of your body.

Now direct your attention to your toes. Feel as though a warm bath surrounds and interpenetrates your toes. It calms and sooths. Feel this part of your body going to a deep state of relaxation. Feel the relaxation spreading to nearby parts of your body.

Now direct your attention to the soles of your feet. Feel as though a warm bath surrounds and interpenetrates the soles of your feet. It calms and sooths. Feel this part of your body going to a deep state of relaxation. Feel the relaxation spreading to nearby parts of your body.

Mentally repeat and visualize the number three. Three. Three. Three. Three. Three.

You are now at level three, the level of complete physical relaxation.

Now remember something that makes you feel safe, calm and relaxed. This may be a favorite meadow, or a mountain trail, or a beach, or just being with someone special. Remember how it makes you feel. You are safe and happy.

You feel emotionally relaxed. Nothing can disturb you at this level of the mind.

Mentally repeat and visualize the number two. Two. Two. Two. Two. Two.

You are now at level two. The level of complete emotional relaxation.

Mentally repeat and visualize the number one. One. One. One. One. One.

Relax your mind. There is no further need to think. Stop thinking until you are told to come out of levels. Your mind is completely relaxed.

This is level one, the level of complete and healthy relaxation.

To come out of levels, you may count from one to five. When you count the number two, you will be emotionally relaxed, but thinking clearly. When you count three, you will be emotionally healthy and active. When you count five, you will be wide-awake and healthy, ready to continue your day.

You will also come out of levels if someone taps you on the shoulder three times.

Now I'm going to count from one to five. As I count, you are going to slowly come up in levels.

One. Two. Coming up slowly.

Three. At the count of five, you will be back to your normal waking level, wide awake and full of energy.

Four. Five. (Snap fingers) Eyes open. You are wide awake, feeling great.

METHOD 2 [105]

This has no tension and includes an introduction to your etheric body. Your etheric body is literally your 'second body'. It usually coincides with your physical body, but is subtler, being made up of etheric material.

The audio recording of this method is 22 minutes long.

After you have practiced this method for a week or two and at least several times, you should graduate to Method 3.

TRANSCRIPT OF METHOD 2

Make yourself comfortable. Direct your eyes upward at an angle of 45 degrees. Keep your head straight and focus on one spot.

Mentally repeat and visualize the number five. Five. Five. Five. Five. Five. This is level five, your normal waking level of mind.

Close your eyes. Keep your eyes closed until I tell you to open them again.

Mentally repeat and visualize the number four. Four. Four. Four. Four. Four. You are now at level four.

Now direct your attention to your scalp. Feel as though a warm bath surrounds and interpenetrates your scalp. It calms and sooths. Feel this part of your body going to a deep state of relaxation. Feel the relaxation spreading to nearby parts of your body.

---

[105] http://www.awinsystem.org/Handouts/MedMethod2.mp3

Now direct your attention to your forehead. Feel as though a warm bath surrounds and interpenetrates your forehead. It calms and sooths. Feel this part of your body going to a deep state of relaxation. Feel the relaxation spreading to nearby parts of your body.

Now direct your attention to your eyelids. Feel as though a warm bath surrounds and interpenetrates your eyelids. It calms and sooths. Feel this part of your body going to a deep state of relaxation. Feel the relaxation spreading to nearby parts of your body.

Now direct your attention to your face. Feel as though a warm bath surrounds and interpenetrates your face. It calms and sooths. Feel this part of your body going to a deep state of relaxation. Feel the relaxation spreading to nearby parts of your body.

Now direct your attention to your neck. Feel as though a warm bath surrounds and interpenetrates your neck. It calms and sooths. Feel this part of your body going to a deep state of relaxation. Feel the relaxation spreading to nearby parts of your body.

Now direct your attention to your shoulders. Feel as though a warm bath surrounds and interpenetrates your shoulders. It calms and sooths. Feel this part of your body going to a deep state of relaxation. Feel the relaxation spreading to nearby parts of your body.

Now direct your attention to your chest. Feel as though a warm bath surrounds and interpenetrates your chest. It calms and sooths. Feel this part of your body going to a deep state of relaxation. Feel the relaxation spreading to nearby parts of your body.

Now direct your attention to the organs within your chest area. Feel as though a warm bath surrounds and interpenetrates your torso. It calms and sooths. Feel this part of your body going to a deep state of relaxation. Feel the relaxation spreading to nearby parts of your body.

Now direct your attention to your back. Feel as though a warm bath surrounds and interpenetrates your back. It calms and sooths. Feel this part of your body going to a deep state of relaxation. Feel the relaxation spreading to nearby parts of your body.

Now direct your attention to your abdomen. Feel as though a warm bath surrounds and interpenetrates your abdomen. It calms and sooths. Feel this part of your body going to a deep state of relaxation. Feel the relaxation spreading to nearby parts of your body.

Now direct your attention to your buttocks. Feel as though a warm bath surrounds and interpenetrates your buttocks. It calms and sooths. Feel this part of your body going to a deep state of relaxation. Feel the relaxation spreading to nearby parts of your body.

Now direct your attention to your thighs. Feel as though a warm bath surrounds and interpenetrates your thighs. It calms and sooths. Feel this part of your body going to a deep state of relaxation. Feel the relaxation spreading to nearby parts of your body.

Now direct your attention to your knees. Feel as though a warm bath surrounds and interpenetrates your knees. It calms and sooths. Feel this part of your body going to a deep state of relaxation. Feel the relaxation spreading to nearby parts of your body.

Now direct your attention to your calves. Feel as though a warm bath surrounds and interpenetrates your calves. It calms and sooths. Feel this part of your body going to a deep state of relaxation. Feel the relaxation spreading to nearby parts of your body.

Now direct your attention to your toes. Feel as though a warm bath surrounds and interpenetrates your toes. It calms and sooths. Feel this part of your body going to a deep state of relaxation. Feel the relaxation spreading to nearby parts of your body.

Now direct your attention to the soles of your feet. Feel as though a warm bath surrounds and interpenetrates the soles of your feet. It calms and sooths. Feel this part of your body going to a deep state of relaxation. Feel the relaxation spreading to nearby parts of your body.

Mentally repeat and visualize the number three. Three. Three. Three. Three. Three.

You are now at level three, the level of complete physical relaxation.

Now remember something that makes you feel safe, calm and relaxed. This may be a favorite meadow, or a mountain trail, or a beach, or just being with someone special. Remember how it makes you feel. You are safe and happy. You feel emotionally relaxed. Nothing can disturb you at this level of the mind.

Mentally repeat and visualize the number two. Two. Two. Two. Two. Two.

You are now at level two. The level of complete emotional relaxation.

Mentally repeat and visualize the number one. One. One. One. One. One.

Relax your mind. There is no further need to think. Stop thinking until you are told to come out of levels. Your mind is completely relaxed.

This is level one, the level of complete and healthy relaxation.

To come out of levels, you may count from one to five. When you count the number two, you will be emotionally relaxed, but thinking clearly. When you count three, you will be emotionally healthy and active. When you count five, you will be wide-awake and healthy, ready to continue your day.

You will also come out of levels if someone taps you on the shoulder three times.

Your physical body is completely relaxed. Keep your physical body relaxed while you allow your etheric body to move a little.

Now make your etheric body coincide with your physical body. Just imagine and visualize that your two bodies are the same shape, size and place.

Now shift your etheric body six inches to the right. Remember how that feels.

Now shift your etheric body back to coincide with your physical body. Remember how that feels.

Now shift your etheric body six inches to the left. Remember how that feels.

Now shift your etheric body back to coincide with your physical body.

Now I'm going bring you out of levels by counting from one to five. As I count, you are going to slowly come up in levels.

One. Two. Coming up slowly.

Three. At the count of five, you will be back to your normal waking level, wide awake and full of energy.

Four. Five. (Snap fingers) Eyes open. You are wide awake, feeling great.

METHOD 3 [106]

This method is shorter and allows you to come up in levels when you choose. It includes a little more about your etheric body. You should learn to practice meditation without the recording. When you practice without the recording, don't hurry, but learn to relax in a shorter time. After a few months of regular practice, you should be able to go into meditation in less than a minute. After you practice the Comet Exercise and astral travel using this technique for a while, you may find that meditation Method 4 will work for you.

TRANSCRIPT OF METHOD 3

Mentally repeat and visualize the number five. Five. Five. Five. Five. Five. This is level five, your normal waking level of mind.

Close your eyes.

---

[106] http://www.awinsystem.org/Handouts/MedMethod3.mp3

Mentally repeat and visualize the number four. Four. Four. Four. Four. Four. You are now at level four.

Now direct your attention to your scalp. Relax your scalp. Feel this part of your body going to a deep state of relaxation.

Now direct your attention to your forehead. Relax your forehead. Feel this part of your body going to a deep state of relaxation.

Now direct your attention to your eyelids. Relax your eyelids. Feel this part of your body going to a deep state of relaxation.

Now direct your attention to your face. Relax your face. Feel this part of your body going to a deep state of relaxation.

Now direct your attention to your neck. Relax your neck. Feel this part of your body going to a deep state of relaxation.

Now direct your attention to your shoulders. Relax your shoulders. Feel this part of your body going to a deep state of relaxation.

Now direct your attention to your chest. Relax your chest. Feel this part of your body going to a deep state of relaxation.

Now direct your attention to the organs within your chest area. Feel those organs functioning in a healthy and relaxed manner..

Now direct your attention to your back. Relax your back. Feel this part of your body going to a deep state of relaxation.

Now direct your attention to your abdomen. Relax your abdomen. Feel this part of your body going to a deep state of relaxation.

Now direct your attention to your buttocks. Relax your buttocks. Feel this part of your body going to a deep state of relaxation.

Now direct your attention to your thighs. Relax your thighs. Feel this part of your body going to a deep state of relaxation.

Now direct your attention to your knees. Relax your knees. Feel this part of your body going to a deep state of relaxation.

Now direct your attention to your calves. Relax your calves. Feel this part of your body going to a deep state of relaxation.

Now direct your attention to your toes. Relax your toes. Feel this part of your body going to a deep state of relaxation.

Now direct your attention to the soles of your feet. Relax the soles of your feet. Feel this part of your body going to a deep state of relaxation.

Mentally repeat and visualize the number three. Three. Three. Three. Three. Three.

You are now at level three.

Now remember something that makes you feel safe, calm and relaxed.

Mentally repeat and visualize the number two. Two. Two. Two. Two. Two.

You are now at level two.

Mentally repeat and visualize the number one. One. One. One. One. One.

You are now at level one.

Keep your physical body completely relaxed.

Lift your etheric right leg. Remember how that feels. Put your etheric right leg back with your physical right leg.

Lift your etheric left leg. Remember how that feels. Put your etheric left leg back with your physical left leg.

Now sit forward in your etheric body. Be aware in your etheric body. Look around. Remember what you perceive. Sit back where you were.

Stay at this level for as long as you wish. When you are ready, you can come out of levels by counting from one to five.

METHOD 4

This is an advanced meditation method that takes you directly into astral travel. To be effective, you first need to practice methods 1 through 3 for several months, practice the Comet exercise, and practice astral travel using method 3 without the CD.

TRANSCRIPT OF METHOD 4

Close your eyes. Visualize a plane moving down from just above your head. Each part of your body relaxes as you move the plain down.

You move the plane at your own pace, and you can start over if you need to. When the plane moves below your feet, you will be relaxed.

Now move another plane down from just above your head. (This might be thought of like you are moving up into water, and the plane is the lower surface of the water.)

This time, as you move the plane lower, you become aware of the parts of your etheric body above the plane (as if they were in the water), and they become free to move as you choose.

When the plane moves below your feet, you can be aware in your entire etheric body and move it where you will.

Appendix E: Etheric Vision Exercises

Everyone is subconsciously aware of the etheric level of reality, but we are taught from early childhood that such perceptions are imagination or make-believe. This is unfortunate because life can be much richer and more effective when a person has awareness of the etheric, astral, and higher levels of reality.

The AWIN system guides a person through the development of awareness of other levels. Accuracy is enhanced by following the system, but feedback from one or more teachers and other students is highly recommended. The use of imagination is good, but only when you can tell it apart from reality. Always check the accuracy of your perceptions whenever you can do so.

The integration of personal consciousness is a somewhat more advanced topic. Human consciousness has multiple aspects or parts. These include the subconscious, the intellect, the emotions, and the Higher Self, to name a few.

As a psychiatrist would tell you, these various aspects of your consciousness can work together or can work against each other. A person's life is often very difficult when major aspects of his or her consciousness do not work well together.

By integrating your consciousness, the various aspects of your consciousness tend to work well together, resulting in a smoother and more effective life. In fact, by integrating your subconscious, intellect, and emotions, you can effectively boost your intelligence dramatically.

The AWIN system provides principles and techniques to integrate a person's consciousness. These principles and techniques are taught at the intermediate and advanced levels.

COMET EXERCISE

This is an individual exercise.

PURPOSE

The Comet Exercise [107] is useful in at least two ways. First, it gives experience in sensing and controlling energy. Second, the aura seems to be strengthened by this exercise. If you become proficient with the Comet Exercise, the techniques translate directly to scientific observation.

PRELIMINARIES

The Comet Exercise is based on the AWIN meditation methods. You should have practiced Method 3 at least three times.

METHOD

Enter meditation using a method similar to Method 3. Choose a place in the physical body, preferably away from significant chakras, if you know what those are. A good place to start is in an arm or leg muscle. Visualize an empty ball of energy there. Move the ball slowly, and as it moves, it gains energy. The ball should move about two feet in about ten seconds. Move the

---

[107] http://www.awinsystem.org/Handouts/Comet.pdf

ball only within your physical body, visualizing the ball growing slowly in intensity (not in size). You do not want to take too much energy from any one part of your body, so move the ball in a random pattern, maintaining a uniform slow speed. Continue to move the ball randomly, increasing in intensity, for as long as you wish. Then continue moving the ball, but visualize it decreasing in intensity. When the ball is no longer perceptible, you have finished the exercise.

CAUTIONS

The ball is composed of your energy and should not be influenced by anything other than you. There may be dangers if you allow the ball to move outside your physical body. The energy in the ball is taken from your etheric body or from your aura. If that energy becomes unbalanced (or too weak or too strong) in any chakra, strange things might happen to your physical and psychological health (so some people have written). Always move the ball throughout your body, not taking too much energy from any area and always distributing the energy widely. Do not become obsessed by this aspect of the exercise. Instead, use random movements of the ball, not allowing repetitive patterns.

HUMAN AURA EXERCISE

This exercise is intended to be used by a group, such as a class who are developing etheric vision together.

PURPOSE

The purpose of this exercise is to gain experience perceiving human auras, with feedback to increase accuracy.

METHOD

Select one member, referred to as the actor, to manipulate his or her aura. The other participants are then observers. All participants go into meditation together, in the same room, using a method similar to meditation Method 3. When all are deep in meditation, a signal is given and the actor significantly alters the size, shape, or position of his or her aura. If possible, the actor provides between 1 and 3 reference points by indicating when the observers should make note of their perceptions. If the signal is verbal, it should be no more informative than, "Remember your perceptions." It may be better to use a device that makes a small, but easily recognized sound when a button is pressed. This reduces the physical activity of the actor, and increases his or her ability to maintain deep meditation.

When finished, a signal should be given for the participants to come out of meditation. After all participants have written a journal of their experiences, including what they perceived at each reference point, the actor describes what he or she intended to do with his or her aura at each reference point. Each observer describes what he or she perceived at each reference point.

Look for consistency between descriptions. The actor probably achieved some degree of success in changing his or her aura, but maybe not. Some of the observers will be more successful in perceiving the actor's aura than others. If two or more participants agree in their descriptions, the likelihood of the accuracy of their perceptions is greater. The idea is to learn from the feedback you get from each other.

CAUTIONS

If too much information or hints are shared prematurely, your perceptions may be biased, resulting in less effectiveness later in your scientific observations. Guided meditation is the least effective approach, where a narrator explains what you should perceive at each step. I recommend that you stick with the method as described, avoiding all hints prior to or during journaling.

PLANT AURA EXERCISE

This exercise can be used by an individual or a group.

PURPOSE

The purpose of this exercise is to gain experience perceiving auras, with feedback to increase accuracy.

METHOD

Enlist the aid of an assistant. Your assistant need not have etheric vision and is not required to meditate or learn with you. He or she is only required to follow instructions while you are meditating.

Select a room at least 10 feet by 10 feet.

Select 2 or 3 live plants roughly one foot tall. The idea is that your assistant will place these plants around the room while you are meditating, and without making any sound that might indicate where the plants are.

Put on a blindfold and use earplugs to reduce the chances of getting physical queues of the location or activity of your assistant. Sit down in the middle of the room and enter meditation. Do not try to perceive anything yet. Your assistant should place the plants randomly around the room, probably different places each time. If significant noise is made by the plant or pot while it is being placed, the assistant must immediately move the plant somewhere else without interacting with you. When finished, the assistant should notify you to begin your observation, probably by a light touch on your arm or hand.

Try to perceive the auras of the plants. Remember what you perceive. Come out of meditation. Make notes in a journal about what you perceived and where you perceived it. Make note of where the plants actually were, relative to what you perceived.

Appendix F: Sample Record Sheet

Here is a sample record sheet for use with the simple sun-sign experiment. If I were the experimenter, I might use such a spreadsheet to make note of the most relevant details of an experiment.

Subject	Target	Birth Date	Birth Time	Birth Loc	Group	A	B	C	Correct Choice	Actual Session Choice Date/Time
Sam	Larry	1/10/1955			P	11	9	10		
	Pam	6/4/1972	14:31		P	3	4	2		
	Ben	8/22/1968			E	6	5	4		
Alicia	Mary	5/21/1968			C	8	9	10		
	Lidia	10/9/1979			P	7	8	6		

NOTES

The Subject and Target fields may be their names or any identifiers.

The birth date is required. The birth time and location are desirable.

The Group is "P" for primary, "E" or excluded or "C" C for control.

Columns A, B, and C are the 3 personality types in randomized order. Each column contains the number of the corresponding personality type from Appendix A.

The letters A, B and C can be used to label the personality types when they are shown to the Subjects.

The meaning of the "Correct" choice depends on the group. If the Target is in the primary or excluded group, the choice is calculated—the sun-sign. If a member of the control group, this is randomly selected from the three randomly selected personality types.

The "Actual" choice is the one selected by the Subject.

The Session Time/Date may be of interest later.

COMMENTS

The birth time and location are useful to an astrologer, and may help exclude Targets that could be confusing to their associated Subjects. This data is good to know.

The letters A, B and C are used to create indirection so that the Assistants and the Subjects will not be able to guess the choice made by the Experimenter, i.e. to make the experiment double-blind.

A natal chart will be needed for each Target, but only the sun-sign need be noted here.

However, a couple of additional columns could be useful here.

For example, an astrologer may estimate the probability of selecting each choice for a Target in the excluded group, or even some of those in the primary group. This information might be helpful in learning more about and improving our understanding of astrological influences.

About The Author

Jack Bosworth was born and grew up in Michigan. He received a PhD in Computer Science from the University of Michigan in 1975.

The summer before starting college, Jack had an experience that contradicted modern science as he knew it, and he made a commitment to investigate, consistent with his understanding of the scientific method.

He did not see an opportunity to follow up on his commitment until he was in graduate school. He took a class on ESP, followed by a lot of independent study. He has been investigating and learning ever since.

Science Restored is effectively a primer, a starting point for those who have focused entirely on the physical level of reality. Jack has learned a great deal about human consciousness and how our consciousness is related to our environment.

He intends to publish more advanced information in later books.

www.ingramcontent.com/pod-product-compliance
Lightning Source LLC
Chambersburg PA
CBHW060329200326
41519CB00011BA/1884

* 9 7 8 1 9 4 7 5 1 4 0 3 4 *